T0226202

Exercises in Programming Style

Exercises in Programming Style

Second Edition

Cristina Videira Lopes

CRC Press
Taylor & Francis Group
Boca Raton London New York

CRC Press is an imprint of the
Taylor & Francis Group, an **informa** business

A CHAPMAN & HALL BOOK

Second edition published 2021
by CRC Press
6000 Broken Sound Parkway NW, Suite 300, Boca Raton, FL 33487-2742

and by CRC Press
4 Park Square, Milton Park, Abingdon, Oxon OX14 4RN

© 2021 Taylor & Francis Group, LLC
CRC Press is an imprint of Taylor & Francis Group, an Informa business

Library of Congress Cataloging-in-Publication Data

Names: Lopes, Cristina Videira, author.
Title: Exercises in programming style / Cristina Videira Lopes.
Description: Second edition. | Boca Raton : CRC Press, 2020. | Includes bibliographical references and index.
Identifiers: LCCN 2020014298 | ISBN 9780367350208 (paperback) | ISBN 9780367360207 (hardback) | ISBN 9780429343216 (ebook)
Subjects: LCSH: Computer programming.
Classification: LCC QA76.6 .L636 2020 | DDC 005.13--dc23
LC record available at https://lccn.loc.gov/2020014298

ISBN-13: 978-0-367-36020-7 (hbk)
ISBN-13: 978-0-367-35020-8 (pbk)
ISBN-13: 978-0-429-34321-6 (ebk)

Typeset in CMR
by Nova Techset Private Limited, Bengaluru & Chennai, India

To Julia

Contents

Preface to the Second Edition

In the six years since the first edition of this book was published, two things happened that made me want to update the book. The first one was the wide adoption of Python 3. The original edition had all its code in Python 2, which has now reached its official end of life. This second edition updates all the code to Python 3.

But the second, and most important, thing that happened since 2014 was the dizzying developments in machine learning, more specifically, in neural networks (NNs). By 2018, I felt it was my duty, and personal challenge, to capture the basic programming concepts in neural networks in exactly the same way I had done for all other concepts: by doing term frequency in neural networks. This led me into a fascinating tour of that field, focused, as I was, on exploring it using a problem for which neural networks are not typically used – term frequency is a well-specified problem for which we know the exact logic. This second edition includes a whole new part, Part X, featuring several basic programming concepts in NNs.

In the process of doing this tour of neural networks for the term frequency problem, four things became clear. **First**, I had to break the problem down into its smaller components, and show how to solve those smaller sub-problems using NNs. This is because the solution of the complete problem is essentially a pipeline of functions that requires knowledge of too many NN concepts at once. **Second**, although *learning* is the magic sauce for which neural networks are popular, I found myself being even more fascinated by the concept of networks as computing machines. As much as I admire the power of statistics for making predictions based on existing data, the computer engineer in me absolutely wants to program these networks by hand by setting the weights manually. I couldn't help it! Part X is full of neural networks that are programmed manually, without learning. **Third**, the most popular framework for programming NNs, TensorFlow, uses *array programming* concepts at its core. This is not surprising, given that we are dealing, essentially, with linear algebra operations. I realized the first edition had missed this historically important programming style – array programming – so I added a new chapter about it in Part I, Historical. (Hello, APL, apologies for having missed you the first time!) Finally, the **fourth** thing I realized was that I could easily write an entire new book just covering neural network programming concepts. I had to stop myself at six chapters in Part X, but these chapters don't even begin to cover the immense and fertile field of programming ideas in neural networks.

NNs require a fundamentally different way of thinking about computing that is at the same time very low level and very powerful. I am now convinced that every programmer needs to learn this connectionist computing model, not just for the hype that its applications currently enjoy, but in spite of it.

Pierre Baldi was instrumental in my developing an interest in neural networks, and in my ability to navigate that field as an outsider. I thank him for the many conversations we had about all that is covered in Part X. In these past six years, my daughter Julia grew up, and, at points, helped me stay focused on finishing this second edition of the book. I thank her for that. Thank you, also, to my Department Chair, André van der Hoek, and to my Dean, Marios Papaefthymiou, for letting me go on sabbatical in 2018. That allowed me to dive into the world of machine learning. Finally, I want to thank the hundreds of students who have taken my course, and who have enthusiastically provided all sorts of feedback.

Cristina Videira Lopes
Irvine, February 29, 2020

Preface to the First Edition

THE CODE

This book is a companion text for code that is publicly available at
http://github.com/crista/exercises-in-programming-style

WHO WILL BENEFIT FROM THIS BOOK

The collection of code that is the foundation of this book is for everyone who enjoys the art of programming. I've written this book in order to complement and explain the raw code, as some of the idioms may not be obvious. Software developers with many years of experience may enjoy revisiting familiar programming styles in the broad context of this book and learning about styles that may not be part of their normal repertoire.

This book can be used as a textbook for advanced programming courses in computer science and software engineering. Additional teaching materials, such as lecture slides, are also available. The book is not designed for introductory programming courses; it is important for students to be able to crawl (i.e. learn to program under the illusion that there's only one way of doing things) before they can run (i.e. realize that there's a lot more variety). I expect that many of the readers will be students in their junior/senior years or in their early stages of graduate study. The exercise list at the end of each chapter is a good mechanism for testing the reader's understanding of each style. The suggested further readings are more appropriate for graduate students.

This book may also be of interest to writers, especially those who know a little programming or have a strong interest in programming technology. Despite important differences, there are many similarities between writing programs and writing in general.

MOTIVATION FOR THESE EXERCISES

In the 1940s, the French writer Raymond Queneau wrote a jewel of a book called *Exercises in Style*, featuring 99 renditions of the exact same story, each written in a different style. The book is a masterpiece of writing technique, as it illustrates the many different ways a story can be told. The story being fairly trivial and always the same, highlights form, rather than content; it illustrates how the decisions we make in telling a story affect the perception of that story.

Queneau's story is trivially simple and can be told in two sentences: The narrator is on the "S" bus and notices a man with a long neck who is wearing a hat, and who gets into an altercation with the man sitting next to him. Two hours later, the narrator sees this same man near the Saint Lazare train station, with a friend, and the friend is giving this man some advice regarding an extra button on his overcoat. That's it! He then goes through 99 renditions of this story using, for example, litotes, metaphors, animism, etc.

Over the years, as an instructor of many programming-intensive courses, I noticed that often students have a hard time understanding the different ways of writing programs and of designing systems, in general. They have been trained in one, at most two, programming languages, so they understand only the styles that are encouraged by those languages, and have a hard time wrapping their heads around other styles. It's not their fault. Looking at the history of programming languages and the lack of pedagogical material on style in most computer science programs, one hardly gets exposed to the issue until after an enormous amount of experience is accumulated. Even then, style is seen as an intangible property of programs that remains elusive to explain to others – and over which many technical arguments ensue. So, in order to give programming styles the proper due, and inspired by Queneau, I decided to embark on the project of writing the exact same computational task in as many styles as I have come across over the years.

So what is *style*? In Queneau's circle of intellectuals, a group known as *Oulipo* (for French *Ouvroir de la littérature potentielle*, roughly translated as "workshop of potential literature"), style was nothing but the consequence of *creating under constraints*, often based on mathematical concepts such as permutations or lipograms. These constraints are used as a means to create something intellectually interesting besides the story itself. The ideas caught on, and over the years, several literary works have been created using Oulipo's constraints.

In this book, too, programming style is what results from writing programs under a set of constraints. Constraints can come from external sources or they can be self-imposed; they can capture true challenges of the environment or they can be artificial; they can come from past experiences and measurable data or they can come from personal preferences. Independent of their origin, constraints are the seeds of style. By honoring different constraints, we can write a variety of programs that are virtually identical in terms of *what* they do, but that are radically different in terms of *how* they do it.

In the universe of all things a good programmer must know, I see collections of programming styles as being as important as any collection of data structures and algorithms, but with a focus on human effects rather than on computing effects. Programs convey information not just to the computers but, more importantly, to the people who read them. As with any form of expression, the consequences of *what* is being said are shaped and influenced by *how* they are being said. An advanced programmer needs not be able to just write correct programs that perform well; he/she needs to be able to choose appropriate styles for expressing those programs for a variety of purposes.

Traditionally, however, it has been much easier to teach algorithms and data structures than it is to teach the nuances of programming expression. Books on data structures and algorithms all follow more or less the same formula: pseudo-code, explanation, and complexity analysis. The literature on programming tends to fall into two camps: books that explain programming *languages* and books that present collections of design or architectural *patterns*. However, there is a continuum in the spectrum of how to write programs that go from the concepts that the programming languages encourage/enforce to the combination of program elements that end up making up the program; languages and patterns feed on each other, and separating them as two different things creates a false dichotomy. Having come across Queneau's body of work, it seemed to me that his focus on *constraints* as the basis for explaining expression styles was a perfectly good model for unifying a lot of important creative work in the programming world.

I should note that I'm not the first one to look at constraints as a good unifying principle for explaining style in software systems. The work on *architectural styles* has taken that approach for a long time. I confess that the notion that style arises from constraints (some things are disallowed, some things must exist, some things are limited, etc.) was a bit hard to understand at first. After all, who wants to write programs under constraints? It wasn't until I came across Queneau's work that the idea made perfect sense.

Like Queneau's story, the computational task in this book is trivial: given a text file, we want to produce the list of words in the file and their frequencies, and print them out in decreasing order of frequency. This computational task is known as **term frequency**. This book contains 33 different styles for writing the term frequency task, one in each chapter. Unlike Queneau's book, I decided to verbalize the constraints in each style and explain the example programs. Given the target audience, I think it's important to provide those insights explicitly rather than leaving them to the reader's interpretation. Each chapter starts by presenting the constraints of the style, then it shows an example program; a detailed explanation of the code follows; most chapters have additional sections regarding the use of the style in systems design and another section on the historical context in which the programming style emerged. History is important; a discipline should not forget the origins of its core ideas. I hope the readers will be curious enough to follow through some of the suggested further readings.

Why 33 styles? I chose 33 as a bounded personal challenge. Queneau's book has 99 styles. Had I set my goal to writing a book with 99 chapters, I probably never would have finished it! The public repository of code that is the basis for this book, however, is likely to continue to grow. The styles are grouped into nine categories: historical, basic, function composition, objects and object interactions, reflection and metaprogramming, adversity, data-centric, concurrency, and interactivity. The categories emerged as a way to organize the book, grouping together styles that are more related to each other than to the others. Other categorizations would be possible.

Similar to Queneau's book, these exercises in programming style are exactly that: *exercises*. They are the sketches, or arpeggios, of software; they aren't the music. A piece of real software usually employs a variety of styles for the different parts of the system. Furthermore, all these styles can be mixed and matched, creating hybrids that are interesting in themselves.

Finally, one last important remark. Although Queneau's book was the inspiration for this project, software is not exactly the same as the language arts; there are utility functions attached to software design decisions, i.e. some expressions are better than others for specific objectives.[1] In this book I try to stand clear of judgments of good and bad, except in certain clear cases. It is not up to me to make those judgments, since they depend heavily on the context of each project.

ACKNOWLEDGMENTS

I would like to thank the following people for valuable feedback on earlier drafts of this book: Richard Gabriel, Andrew Black, Guy Steele, James Noble, Paul Steckler, Paul McJones, Laurie Tratt, Tijs van der Storm, and the students of INF 212 / CS 235 (Winter 14) at UC Irvine, especially Matias Giorgio and David Dinh.

Thanks also to members of the IFIP Working Group 2.16, where I first presented the idea of this book, and whose reactions were critical for shaping the material.

A special thanks to the contributors to the exercises-in-style code repository so far: Peter Norvig, Kyle Kingsbury, Sara Triplett, Jørgen Edelbo, Darius Bacon, Eugenia Grabrielova, Kun Hu, Bruce Adams, Krishnan Raman, Matias Giorgio, David Foster, Chad Whitacre, Jeremy MacCabe, and Mircea Lungu.

---

[1]Maybe that's also the case for the language arts, but I'm afraid I don't know enough!

Prologue

TERM FREQUENCY

L IKE QUENEAU'S STORY, the computational task in this book is trivial: given a text file, we want to display the N (e.g. 25) most frequent words and corresponding frequencies ordered by decreasing value of frequency. We should make sure to normalize for capitalization and to ignore stop words like "the," "for," etc. To keep things simple, we don't care about the ordering of words that have equal frequencies. This computational task is known as **term frequency**.

Here is an example of an input file and corresponding output after computing the term frequency:

```
Input:
    White tigers live mostly in India
    Wild lions live mostly in Africa

Output:
    live - 2
    mostly - 2
    africa - 1
    india - 1
    lions - 1
    tigers - 1
    white - 1
    wild - 1
```

If we were to run this flavor of term frequency on Jane Austen's *Pride and Prejudice* available from the Gutenberg Collection, we would get the following output:

```
mr   -   786
elizabeth   -   635
very   -   488
darcy   -   418
such   -   395
mrs   -   343
much   -   329
more   -   327
```

```
bennet   -   323
bingley   -   306
jane   -   295
miss   -   283
one   -   275
know   -   239
before   -   229
herself   -   227
though   -   226
well   -   224
never   -   220
sister   -   218
soon   -   216
think   -   211
now   -   209
time   -   203
good - 201
```

This book's example programs cover this term frequency task. Additionally, all chapters have a list of exercises. One of those exercises is to write another simple computational task using the corresponding style. Some suggestions are given below.

These computational tasks are simple enough for any advanced student to tackle easily. Algorithmic difficulties out of the way, the focus should be on following the constraints that underlie each style.

WORD INDEX

Given a text file, output all words alphabetically, along with the page numbers on which they occur. Ignore all words that occur more than 100 times. Assume that a page is a sequence of 45 lines. For example, given *Pride and Prejudice*, the first few entries of the index would be:

```
abatement - 89
abhorrence - 101, 145, 152, 241, 274, 281
abhorrent - 253
abide - 158, 292
. . .
```

WORDS IN CONTEXT

Given a text file, display certain words alphabetically and in context, along with the page numbers of the pages in which they occur. Assume that a page is a sequence of 45 lines. Assume that context consists of the preceding and succeeding two words. Ignore punctuation. For example, given *Pride and*

Prejudice, the words "concealment" and "hurt" would result in the following output:

perhaps this **concealment** this disguise - 150
purpose of **concealment** for no - 207
pride was **hurt** he suffered - 87
must be **hurt** by such - 95
and are **hurt** if i - 103
pride been **hurt** by my - 145
must be **hurt** by such - 157
infamy was **hurt** and distressed – 248

Suggestion of words for the words in context task: concealment, discontented, hurt, agitation, mortifying, reproach, unexpected, indignation, mistake, and confusion.

PYTHONISMS

The example code used in this book is all written in Python, but expertise in Python is not necessary in order to understand the styles. In fact, one of the exercises in all of the chapters is to write the example program in another language. As such, the reader needs only to be able to *read* Python without needing to *write* in Python.

Python is relatively easy to read. There are, however, a few corners of the language that may confuse readers coming from other languages. I explain some of them here.

- **Lists.** In Python, a list is a primitive data type supported by dedicated syntax that is normally associated with arrays in C-like languages. Here is an example of a list: `mylist = [0, 1, 2, 3, 4, 5]`. Python doesn't have an array as a primitive data type,[2] and most situations that would use an array in C-like languages use a list in Python.

- **Tuples.** A tuple is an immutable list. Tuples are also primitive data types supported by dedicated syntax that is normally associated with lists in Lisp-like languages. Here is an example of a tuple: `mytuple = (0, 1, 2, 3, 4)`. Tuples and lists are handled in similar ways, except for the fact that tuples are immutable, so the operations that change lists don't work on tuples.

- **List indexing.** List and tuple elements are accessed by index like this: `mylist[some_index]`. The lower bound of a list is index 0, like in C-like languages, and the list length is given by `len(mylist)`. Indexing a list can be much more expressive than this simple example suggests. Here are some more examples:

---

[2] There is an array data object, but it's not a primitive type of the language and it doesn't have any special syntax. It's not used as much as lists.

- `mylist[0]` – first element of the list
- `mylist[-1]` – last element of the list
- `mylist[-2]` – next-to-last element of the list
- `mylist[1:]` – a list starting at index 1 until the end of `mylist`
- `mylist[1:3]` – a list starting at index 1 and stopping before index 3 of `mylist`
- `mylist[::2]` – a list containing every other element of `mylist`
- `mylist[start:stop:step]` – a list containing every `step` element between `start` and `stop` indexes of `mylist`

- **Bounds.** Indexing an element beyond the length of a list results in an IndexError. For example, trying to access the 4th element of a list of 3 elements (e.g. [10, 20, 30][3]) results in an IndexError, as expected. However, many Python operations on lists (and collections in general) are constructivist with respect to indexing. For example, obtaining a list consisting of the range from 3 to 100 in a list with only 3 elements (e.g. [10, 20, 30][3:100]) results in an empty list ([]) rather than an IndexError. Similarly, any range that partially covers a list results in whatever part of the list is covered, with no IndexError (e.g. [10, 20, 30][2:10] results in [30]). This constructivist behavior may be puzzling at first for people used to more intolerant languages.

- **Dictionaries.** In Python, a dictionary, or map, is also a primitive data type supported by dedicated syntax. Here is an example of a dictionary: `mydict = {'a' : 1, 'b' : 2}`. This particular dictionary maps two string keys to two integer values; in general, keys and values can be of any type. In Java, these kinds of dictionaries can be found in the form of the `HashMap` class (among others), and in C++ they can be found in the form of the class template `map` (among others).

- **self.** In most object-oriented languages, the reference that an object has to itself is implicitly available through special syntax. For example, `this` in Java and C++, `$this` in PHP, or `@` in Ruby. Unlike these languages, Python has no special syntax for it. Moreover, instance methods are simply class methods that take an object as the first parameter; this first parameter is called `self` by convention, but not by special mandate of the language. Here is an example of a class definition with two instance methods:

```
1   class Example:
2       def set_name(self, n):
3           self._name = n
4       def say_my_name(self):
5           print self._name
```

Both methods have a parameter named `self` in the first position, which is then accessed in their bodies. There is nothing special about the word `self`, and the methods could use any other name, for example `me` or `my` or even `this`, but any word other than `self` will be frowned upon by Python programmers. Calling instance methods, however, may be surprising, because the first parameter is omitted:

```
e = Example()
e.set_my_name(``Heisenberg'')
e.say_my_name()
```

This mismatch on the number of parameters is due to the fact that the dot-notation in Python ('.') is simply syntactic sugar for this other, more primitive form of calling the methods:

```
e = Example()
Example.set_my_name(e,  ``Heisenberg'')
Example.say_my_name(e)
```

- **Constructors.** In Python, a constructor is a regular method with the name __init__ (two underscores on each side of the word). Methods with this exact name are called automatically by the Python runtime right after object creation. Here is one example of a class with a constructor, and its use:

```
1    class Example:
2        # This is the constructor of this class
3        def __init__(self, n):
4            self._name = n
5        def say_my_name(self):
6            print self._name
7
8    e = Example(``Heisenberg'')
9    e.say_my_name()
```

Author

Cristina (Crista) Videira Lopes is a Professor of Software Engineering at
the Donald Bren School of Information and Computer Sciences, University of
California, Irvine. Her research focuses on software engineering for large-scale
data and systems. Early in her career, she was a founding member of the team
at Xerox PARC that developed Aspect-Oriented Programming and AspectJ.
Along with her research program, she is also a prolific software developer. Her
open source contributions include acoustic software modems, and the virtual
world server OpenSimulator. She is a co-founder of a company specializing in
online virtual reality for early-stage sustainable urban redevelopment projects.
She developed and maintains a search engine for OpenSimulator-based virtual
worlds.

Dr. Lopes has a PhD from Northeastern University, and MS and BS degrees
from Instituto Superior Técnico in Portugal. She is the recipient of several
National Science Foundation grants, including a prestigious CAREER Award.
She is an ACM Distinguished Scientist and an IEEE Fellow.

I

Historical

Computing systems are like an onion, with layers upon layers of abstraction developed over the years in order to facilitate the expression of intent. It is important to know what the inner layers really entail. The first three programming styles illustrate what programming was like several decades ago, and to some extent, what it still is – because ideas keep getting reinvented.

## CHAPTER 1

Good Old Times

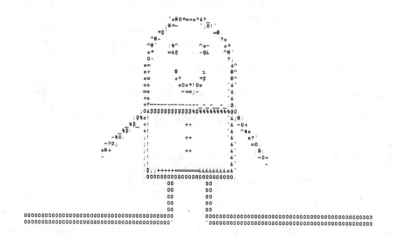

1.1 CONSTRAINTS

▷ Very small amount of primary memory, typically orders of magnitude smaller than the data that needs to be processed/generated.

▷ No identifiers – i.e. no variable names or tagged memory addresses. All we have is memory that is addressable with numbers.

1.2 A PROGRAM IN THIS STYLE

```python
1  #!/usr/bin/env python
2  import sys, os, string
3
4  # Utility for handling the intermediate 'secondary memory'
5  def touchopen(filename, *args, **kwargs):
6      try:
7          os.remove(filename)
8      except OSError:
9          pass
10     open(filename, "a").close() # "touch" file
11     return open(filename, *args, **kwargs)
12
13 # The constrained memory should have no more than 1024 cells
14 data = []
15 # We're lucky:
16 # The stop words are only 556 characters and the lines are all
17 # less than 80 characters, so we can use that knowledge to
18 # simplify the problem: we can have the stop words loaded in
19 # memory while processing one line of the input at a time.
20 # If these two assumptions didn't hold, the algorithm would
21 # need to be changed considerably.
22
23 # Overall strategy: (PART 1) read the input file, count the
24 # words, increment/store counts in secondary memory (a file)
25 # (PART 2) find the 25 most frequent words in secondary memory
26
27 # PART 1:
28 # - read the input file one line at a time
29 # - filter the characters, normalize to lower case
30 # - identify words, increment corresponding counts in file
31
32 # Load the list of stop words
33 f = open('../stop_words.txt')
34 data = [f.read(1024).split(',')] # data[0] holds the stop words
35 f.close()
36
37 data.append([])       # data[1] is line (max 80 characters)
38 data.append(None)     # data[2] is index of the start_char of word
39 data.append(0)        # data[3] is index on characters, i = 0
40 data.append(False)    # data[4] is flag indicating if word was found
41 data.append('')       # data[5] is the word
42 data.append('')       # data[6] is word,NNNN
43 data.append(0)        # data[7] is frequency
44
45 # Open the secondary memory
46 word_freqs = touchopen('word_freqs', 'rb+')
47 # Open the input file
48 f = open(sys.argv[1], 'r')
49 # Loop over input file's lines
50 while True:
51     data[1] = [f.readline()]
52     if data[1] == ['']: # end of input file
53         break
```

```
54    if data[1][0][len(data[1][0])-1] != '\n': # If it does not end
          with \n
55        data[1][0] = data[1][0] + '\n' # Add \n
56    data[2] = None
57    data[3] = 0
58    # Loop over characters in the line
59    for c in data[1][0]: # elimination of symbol c is exercise
60        if data[2] == None:
61            if c.isalnum():
62                # We found the start of a word
63                data[2] = data[3]
64        else:
65            if not c.isalnum():
66                # We found the end of a word. Process it
67                data[4] = False
68                data[5] = data[1][0][data[2]:data[3]].lower()
69                # Ignore words with len < 2, and stop words
70                if len(data[5]) >= 2 and data[5] not in data[0]:
71                    # Let's see if it already exists
72                    while True:
73                        data[6] = str(word_freqs.readline().strip
                              (), 'utf-8')
74                        if data[6] == '':
75                            break;
76                        data[7] = int(data[6].split(',')[1])
77                        # word, no white space
78                        data[6] = data[6].split(',')[0].strip()
79                        if data[5] == data[6]:
80                            data[7] += 1
81                            data[4] = True
82                            break
83                    if not data[4]:
84                        word_freqs.seek(0, 1) # Needed in Windows
85                        word_freqs.write(bytes("%20s,%04d\n" % (
                              data[5], 1), 'utf-8'))
86                    else:
87                        word_freqs.seek(-26, 1)
88                        word_freqs.write(bytes("%20s,%04d\n" % (
                              data[5], data[7]), 'utf-8'))
89                    word_freqs.seek(0,0)
90                # Let's reset
91                data[2] = None
92        data[3] += 1
93  # We're done with the input file
94  f.close()
95  word_freqs.flush()
96
97  # PART 2
98  # Now we need to find the 25 most frequently occurring words.
99  # We don't need anything from the previous values in memory
100 del data[:]
101
102 # Let's use the first 25 entries for the top 25 words
103 data = data + [[]]*(25 - len(data))
104 data.append('') # data[25] is word,freq from file
105 data.append(0)  # data[26] is freq
106
```

```
107  # Loop over secondary memory file
108  while True:
109      data[25] = str(word_freqs.readline().strip(), 'utf-8')
110      if data[25] == '': # EOF
111          break
112      data[26] = int(data[25].split(',')[1]) # Read it as integer
113      data[25] = data[25].split(',')[0].strip() # word
114      # Check if this word has more counts than the ones in memory
115      for i in range(25): # elimination of symbol i is exercise
116          if data[i] == [] or data[i][1] < data[26]:
117              data.insert(i, [data[25], data[26]])
118              del data[26] #  delete the last element
119              break
120
121  for tf in data[0:25]: # elimination of symbol tf is exercise
122      if len(tf) == 2:
123          print(tf[0], '-', tf[1])
124  # We're done
125  word_freqs.close()
```

Note: If not familiar with Python, please refer to the Prologue (Pythonisms) for an explanation of lists, indexes, and bounds.

1.3   COMMENTARY

I N THIS STYLE, the program reflects the constrained computing environment where it executes. The memory limitations force the programmer to come up with ways of rotating data through the available memory, adding complexity to the computational task at hand. Additionally, the absence of identifiers results in programs where the natural terminology of the problem is absent from the program text, and, instead, is added through comments and documentation. This is what programming was all about in the early 1950s. This style of programming, however, is not extinct; it is still in use today when dealing directly with hardware and when optimizing the use of memory.

The example program may look quite foreign to programmers not used to these kinds of constraints. While this is certainly a program that one doesn't associate with Python or with any of the modern programming languages, it embodies the theme of this book quite well: programming styles emerge from *constraints*. Very often, the constraints are imposed externally – maybe the hardware has limited memory, maybe the assembly language doesn't support identifiers, maybe performance is critical and one must deal directly with the machine, etc.; other times the constraints are self-imposed: the programmer, or the entire development team, decides to adhere to certain ways of thinking about the problems and of writing the code, for many different reasons – maintainability, readability, extensibility, adequacy for the problem domain, past experiences on the part of the developers; or simply, as is the case here, to teach what low-level programming looks like without having to learn new syntax. Indeed, it is possible to write low-level, Good Old Times style programs in just about any programming language!

Having explained the reason for this unusual implementation of term frequency, let's dive into this program. The memory limitations are such that we can't ignore the size of the data to be processed. In the example, we have self-imposed a size of 1024 memory cells (line #15). The term "memory cell" is used here in a somewhat fuzzy manner to denote, roughly, a piece of simple data, such as a character or a number. Given that books like *Pride and Prejudice* contain much more than 1024 characters, we need to come up with ways to read and process the data in small chunks, making heavy use of "secondary memory" (a file) to store the data that doesn't fit in primary memory. Before we start coding, we need to do some back-of-the-envelope calculations about the different options regarding what to hold in primary memory and what to dump to secondary memory, and when (see comments in lines #16 through #26). Then as now, access to primary memory is orders of magnitude faster than access to secondary memory, so these calculations are about optimizing for performance.

Many options could have been pursued, and the reader is encouraged to explore the solution space within this style. The example program is divided into two distinct parts: the first part (lines #28 through #98) processes the input file, counting word occurrences and writing that data into a word-frequency file; the second part (lines #100 through #128) processes the intermediate word-frequency file in order to discover the 25 most frequently occurring words, printing them at the end.

The first part of the program works as follows:

- Hold the stop words, roughly 500 characters, in primary memory (lines #33 through #36)

- Read the input file one line at a time; each line is only 80 characters maximum (lines #50 through #95)

- For each line (lines #60 through #95), filter the characters, identify the words, and normalize them to lowercase

- Retrieve/Write the words and their frequencies from/to secondary memory (lines #73 through #90)

After processing the entire input file like this, we then turn our attention to the word frequencies that have been accumulated in the intermediate file. We need a sorted list of the most frequently occurring words, so the program does the following:

- Keep an ordered list in memory holding the current 25 most frequently occurring words, and their frequencies (line #104)

- Read one line at a time from the file. Each line contains a word and its corresponding frequency (lines #108 through #120)

- If the new word has higher frequency than any of the words in memory, insert it at the appropriate place in the list and remove the word at the end of the list (lines #116 through #120)

- Finally, print the 25 top words and their frequencies (lines #122 through #124) and close the intermediate file (line #126)

As seen, the memory constraint has a strong effect on the algorithm employed, as we must be mindful of how much data there is in memory at any given point in time.

The second self-imposed constraint of this style is the absence of identifiers. This second constraint also has a strong effect on the program, but this effect is of a different nature: readability. There are no variables, as such; there is only a data memory that is accessed by indexing it with numbers. The problem's natural concepts (words, frequencies, counts, sorting, etc.) are completely absent from the program text, and are, instead, indirectly represented as indexes over memory. The only way we can bring those concepts back in is by adding comments that explain what kinds of data the memory cells hold (e.g. see comments in lines #38 through #44 and #103 through #106, among others). When reading through the program, we often need to go back to those comments to remind ourselves what high-level concept a certain memory index corresponds to.

1.4   THIS STYLE IN SYSTEMS DESIGN

In the age of computers with multi-gigabyte RAM, constrained memory such as that shown here is mostly a vague memory from the past. Nevertheless, with modern programming languages that encourage obliviousness with respect to memory management, and with the ever-growing amounts of data that modern programs handle, it is very easy to let memory consumption of programs run out of control, with negative consequences on run-time performance. A certain amount of awareness regarding the consequences that the different programming styles carry for memory usage is always a good thing.

Many applications these days – namely those falling in what's known as *Big Data* – have brought the complexities of dealing with small memory back to the center of attention. In this case, although memory is not scarce in absolute terms, it is much smaller than the size of the data to be processed. For example, if instead of just *Pride and Prejudice* we would apply term frequency to the entire Gutenberg Collection, we would likely not be able to hold all the books in memory at the same time; we might not even be able to hold the list of word frequencies all in memory either. Once the data can't fit in memory all at once, developers must devise smart schemes for (1) representing data so that more of it can fit in memory at any given time; and (2) rotating the data through primary and secondary memory. Programming with these constraints tends to make programs feel like the Good Old Times style.

Regarding the absence of names, one of the main drivers behind programming language evolution throughout the 1950s and 1960s was precisely the elimination of cognitive indirections such as those shown in the example: we want the program texts to reflect the high-level concepts of the domain as much as possible, instead of reflecting the low-level machine concepts and relying on external documentation to do the mapping between the two. But even though programming languages have provided for user-defined named abstractions for a long time, it is not unusual for programmers to fail to name their program elements, Application Programming Interfaces (APIs) and entire components appropriately, resulting in programs, libraries and systems as obscure as the one shown here.

Let this Good Old Times style be a reminder of how thankful we should be for being able to hold so much data in memory and for being able to give appropriate names to each and every one of our program elements!

1.5   HISTORICAL NOTES

This style of programming came directly from the first viable model of computation, the Turing Machine. A Turing Machine consists of an unbounded modifiable state "tape" (the data memory) and a state machine that reads and modifies that state. The Turing Machine had a tremendous influence in the development of computers and how they were programmed. Turing's ideas influenced von Neumann's design of the first computer with stored programs. Turing himself also wrote the specifications of a computing machine known as the Automatic Computing Engine (ACE), which was, in many ways, more advanced than von Neumann's. Because that report was classified by the British government, and also because of the politics following the Second World War, Turing's design was not acted upon until several years later, and still in secrecy. Both von Neumann's architecture and Turing's machines led to the first programming languages in the 1950s, which enforced the concept of programming by reusing and changing state in memory over time.

1.6 FURTHER READING

Bashe, C., Johnson, L., Palmer, J. and Pugh, E. (1986). *IBM's Early Computers: A Technical History* (History of Computing), MIT Press, Cambridge, MA.
  Synopsis: IBM was the major commercial player in the early days of computing machines. This book tells the story of IBM's transition from manufacturer of electromechanical machines to a powerhouse of computing machines.

Carpenter, B.E. and Doran, R.W. (1977). The other Turing Machine. *Computer Journal* 20(3): 269–279.
  Synopsis: An account of one of Turing's technical reports describing a complete architecture for a computing machine based on von Neumann's, but including subroutines, stacks and much more. The original report can be found at
  http://www.npl.co.uk/about/history/notable-individuals/turing/ace-proposal

Turing, A. (1936). On computable numbers, with an application to the Entscheidungs problem. *Proceedings of the London Mathematical Society* 2(42): 230–265.
  Synopsis: The original "Turing Machine." In the context of this book, this paper is suggested not for its mathematics but for the programming model of a Turing Machine: a tape with symbols, a tape reader/writer that moves left and right, and the overwriting of symbols on the tape.

von Neumann, J. (1945). First draft of a report on the EDVAC. Reprinted in *IEEE Annals of the History of Computing* 15(4): 27–43, 1993.
  Synopsis: The original "von Neumann architecture." As with Turing's paper, suggested for the programming model.

1.7 GLOSSARY

Main memory: Often referred to simply as *memory*, this data storage is directly accessible by the CPU. Most data in this storage is volatile in the sense that it does not persist beyond the execution of the programs that use it and also does not persist upon machine power downs. These days, the main memory is random access memory (RAM), meaning that the CPU can address any cell in it quickly, as opposed to having to scan sequentially.

Secondary memory: In contrast to primary memory, secondary memory refers to any storage facility that is not directly accessible by the CPU and that, instead, is indirectly accessed via input/output channels. Data in secondary memory persists in the device through power downs and until it is explicitly deleted. In modern computers, hard disk drives and "pen" drives are the most common secondary storage forms. Access to secondary memory is several orders of magnitude slower than access to primary memory.

1.8   EXERCISES

1.1   *Another language.* Implement the example program in another language, but preserve the style.

1.2   *No identifiers.* The example program still has a few identifiers left, namely in lines #60 (c), #116 (i) and #122 (tf). Change the program so that these identifiers are also eliminated.

1.3   *More lines.* The example program reads one line at a time into memory. In doing so it is underutilizing the main memory. Modify the program so that it loads as many lines as possible into memory without going over the established limit of 1024 memory cells. Justify the number of lines you chose. Check if your version runs faster than the original example program, and explain the result.

1.4   *A different task.* Write one of the tasks proposed in the Prologue using the Good Old Times style.

## CHAPTER 2

Go Forth

2.1 CONSTRAINTS

▷ Existence of a data stack. All operations (conditionals, arithmetic, etc.) are done over data on the stack.

▷ Existence of a heap for storing data that's needed for later operations. The heap data can be associated with names (i.e. variables). As said above, all operations are done over data on the stack, so any heap data that needs to be operated upon needs to be moved first to the stack and eventually back to the heap.

▷ Abstraction in the form of user-defined "procedures" (i.e. names bound to a set of instructions), which may be called something else entirely.

2.2 A PROGRAM IN THIS STYLE

```python
1  #!/usr/bin/env python
2  import sys, re, operator, string
3
4  #
5  # The all-important data stack
6  #
7  stack = []
8
9  #
10 # The heap. Maps names to data (i.e. variables)
11 #
12 heap = {}
13
14 #
15 # The new "words" (procedures) of our program
16 #
17 def read_file():
18     """
19     Takes a path to a file on the stack and places the entire
20     contents of the file back on the stack.
21     """
22     f = open(stack.pop())
23     # Push the result onto the stack
24     stack.append([f.read()])
25     f.close()
26
27 def filter_chars():
28     """
29     Takes data on the stack and places back a copy with all
30     nonalphanumeric chars replaced by white space.
31     """
32     # This is not in style. RE is too high-level, but using it
33     # for doing this fast and short. Push the pattern onto stack
34     stack.append(re.compile('[\W_]+'))
35     # Push the result onto the stack
36     stack.append([stack.pop().sub(' ', stack.pop()[0]).lower()])
37
38 def scan():
39     """
40     Takes a string on the stack and scans for words, placing
41     the list of words back on the stack
42     """
43     # Again, split() is too high-level for this style, but using
44     # it for doing this fast and short. Left as exercise.
45     stack.extend(stack.pop()[0].split())
46
47 def remove_stop_words():
48     """
49     Takes a list of words on the stack and removes stop words.
50     """
51     f = open('../stop_words.txt')
52     stack.append(f.read().split(','))
53     f.close()
54     # add single-letter words
```

```
55    stack[-1].extend(list(string.ascii_lowercase))
56    heap['stop_words'] = stack.pop()
57    # Again, this is too high-level for this style, but using it
58    # for doing this fast and short. Left as exercise.
59    heap['words'] = []
60    while len(stack) > 0:
61        if stack[-1] in heap['stop_words']:
62            stack.pop() # pop it and drop it
63        else:
64            heap['words'].append(stack.pop()) # pop it, store it
65    stack.extend(heap['words']) # Load the words onto the stack
66    del heap['stop_words']; del heap['words'] # Not needed
67
68 def frequencies():
69    """
70    Takes a list of words and returns a dictionary associating
71    words with frequencies of occurrence.
72    """
73    heap['word_freqs'] = {}
74    # A little flavour of the real Forth style here...
75    while len(stack) > 0:
76        # ... but the following line is not in style, because the
77        # naive implementation would be too slow
78        if stack[-1] in heap['word_freqs']:
79            # Increment the frequency, postfix style: f 1 +
80            stack.append(heap['word_freqs'][stack[-1]]) # push f
81            stack.append(1) # push 1
82            stack.append(stack.pop() + stack.pop()) # add
83        else:
84            stack.append(1) # Push 1 in stack[2]
85        # Load the updated freq back onto the heap
86        heap['word_freqs'][stack.pop()] = stack.pop()
87
88    # Push the result onto the stack
89    stack.append(heap['word_freqs'])
90    del heap['word_freqs'] # We don't need this variable anymore
91
92 def sort():
93    # Not in style, left as exercise
94    stack.extend(sorted(stack.pop().items(), key=operator.
          itemgetter(1)))
95
96 # The main function
97 #
98 stack.append(sys.argv[1])
99 read_file(); filter_chars(); scan(); remove_stop_words()
100 frequencies(); sort()
101
102 stack.append(0)
103 # Check stack length against 1, because after we process
104 # the last word there will be one item left
105 while stack[-1] < 25 and len(stack) > 1:
106    heap['i'] = stack.pop()
107    (w, f) = stack.pop(); print(w, '-', f)
108    stack.append(heap['i']); stack.append(1)
109    stack.append(stack.pop() + stack.pop())
```

Note: If not familiar with Python, please refer to the Prologue (Pythonisms) for an explanation of lists, indexes, and bounds.

2.3 COMMENTARY

THIS STYLE takes inspiration in Forth, a small programming language first developed as a personal programming system in the late 1950s by Charles Moore, a programmer working at the time for the Smithsonian Astrophysical Laboratory. This programming system – an interpreter for a simple language – supported the need to handle different equations without having to recompile the program – a time-consuming activity at the time.

This curious little language has at its heart the concept of a stack. Equations are entered in postfix notation, e.g. "3 4 +". Operands are pushed onto the stack one at a time; operators take the operands on the stack and replace them with the result. When data is not immediately needed, it can be placed on a part of the memory called the heap. Besides the stack machine, Forth supports the definition of procedures (called "words" in Forth); these procedures, like the built-in ones, operate over data on the stack.

The syntax of Forth may be hard to understand due to its postfix nature and several special symbols that aren't used in any other language. But the language model is surprisingly simple once we understand the constraints – the stack, the heap, procedures and names. Rather than using a Forth interpreter written in Python, this chapter shows how the constraints underlying Forth can be codified in Python programs, resulting, roughly, in a Forth style of programming. Let's analyze the example program.

To start with, to support the style, we first define the stack (line #7) and the heap (line #12).[1] Next we define a set of procedures ("words" in Forth terminology). These procedures help us divide the problem in smaller sub-steps, such as reading a file (lines #17 through #25), filtering the characters (lines #27 through #36), scanning for words (lines #38 through #45), removing stop words (lines #47 through #66), computing the frequencies (lines #68 through #90), and sorting the result (lines #92 through #94). We'll look into some of these in more detail next. But one thing to notice is that all these procedures use (pop) data from the stack (e.g. lines #22, #36, #45) and end by pushing data onto the stack (e.g. lines #24, #36, #45).

Forth's heap supports the allocation of data blocks that can be – and usually are – bound to names. In other words, variables. The mechanism is relatively low level, as the programmer needs to define the size of the data. In our emulation of Forth's style, we simply use a dictionary (line #12). So for example, in line #56, we are popping the stop words on the stack directly into a variable on the heap named stop_words.

Many parts of the example program are not written in Forth style, but some parts are true to it, so let's focus on those. The procedure

---

[1] In Python, stacks are simply lists; we use them as stacks by invoking the stack operations pop and append (acting as push). Occasionally, extend is also used, as a shortcut to append/push the elements of an entire list onto the stack.

remove_stop_words (starting in line #47), as the name suggests, removes the stop words. When that procedure is called, the stack contains all the words of the input file, properly normalized. The first few words of *Pride and Prejudice* are:

['the', 'project', 'gutenberg', 'ebook', 'of', 'pride', 'and', 'prejudice', ...]

That is how the stack looks like at that point, for that book. Next, we open the stop words file and push the list of stop words onto the stack (lines #51 through #55). To make things simple, we keep them in a list of their own instead of merging them with the rest of the data on the stack. The stack now looks like this:

['the', 'project', 'gutenberg', 'ebook', 'of', 'pride', 'and', 'prejudice', ..., ['a', 'able', 'about', 'across', ...]]

After reading all the stop words from the file and placing them onto the stack, we then pop them out to the heap (line #56), in preparation to process the words of the book that are still on the stack. Lines #60 through #64 iterate through the words on the stack in the following way. Until the stack is empty (test in line #60), we check if the word at the top of the stack (stack[-1] in Python) is in the list of stop words (line #61). In real Forth, this test would be much more low level than what is shown here, as we would need to explicitly iterate through the list of stop words too. In any case, if the word is in the stop words list, we simply pop it out of the stack and ignore it. If the word is not in the list of stop words (line #63), we pop it out of the stack onto another variable in the heap called words – the list accumulates the non-stop words (line #64). When the iteration is over, we take the variable words and place its entire contents back on the stack (line #65). We end up with the stack containing all non-stop words, like this:

['project', 'gutenberg', 'ebook', 'pride', 'prejudice', ...]

At that point, we don't need the variables on the heap anymore, so we discard them (line #66). Forth supports deletion of variables from the heap in the spirit of what is being done here.

The frequencies procedure (starting in line #68) shows one more stylistic detail related to arithmetic operations. That procedure starts with the non-stop words on the stack (as shown above) and ends with placing a dictionary of word frequencies onto the stack (line #89). It works as follows. First, it allocates a variable on the heap called word_freqs that stores the word-frequency pairs (line #73) – it starts with the empty dictionary. It then iterates through the words on the stack. For each word at the top of the stack, it checks whether that word has been seen before (line #78). Again, this test is expressed at a much higher level than it would be in Forth, for performance reasons. If the word has been seen before, we need to increment its frequency count. That is done by pushing the current frequency count onto the stack (line #80), then pushing the value 1 onto the stack (line #81), and then adding those 2 top-most operands on the stack and placing the result on the stack (line #82). If the word had not been seen before (line #83), we simply push the value 1 onto the stack. Finally, we pop both the frequency count (right side of assignment in line #86) and the word itself (left side of assignment in line #86) out of the stack and into the variable on the heap, and move to the next word on top of the stack, until the stack is empty (back to line #75). At the end, as stated before, we push the entire content of the heap variable onto the stack, and delete that variable.

The main function starting in line #98 is the beginning of the program. We start by pushing the name of the input file onto the stack (line #98), and then invoke the procedures sequentially. Note that these procedures are not completely independent of each other: each of them relies on strong assumptions about the data that is left on the stack by the previous one.

Once the counting and sorting is done, we then print out the result (lines #105 through #109). This block of code shows a final stylistic detail related to what Forth calls "indefinite loops," or loops that run until a condition is true. In our case, we want to iterate through the dictionary of word frequencies until we count 25 iterations. So we do the following. We start by pushing the number 0 onto the stack (line #102), on top of the data that is already there (word frequencies), and proceed to an indefinite loop until the top of the stack reaches the number 25. In each iteration, we pop the count out of the stack into a variable (line #106), then pop the next word and frequency out of the stack and print them (line #107); then, we push the count in the variable back to the stack, followed by the value 1, and add them, effectively incrementing the count. The loop, and the program, terminate when the top of the stack has the value 25. The second clause for termination (`len(stack) > 1`) is there in the case of small test files that might not even have 25 words.

Many options could have been pursued, and the reader is encouraged to explore the solution space within this style.

2.4   HISTORICAL NOTES

Early computing machines did not have stacks. The earliest reference for the idea of using a stack in computers can be found in Alan Turing's Automatic Computing Engine (ACE) report in 1945. Unfortunately, that report was classified for many years, so not many knew about it.

Stacks were invented again in the late 1950s by several people independently. It was several more years before computer architectures started to include stacks, and to use them for purposes such as subroutine calls.

Forth was a personal project from a computer maverick that never caught the attention of the dominant players of the time. Forth was entirely done in software, and it has been ported [by Moore] to several generations of computers since 1958. Considering that Moore started using it in the late 1950s, the fact that Forth was a stack machine interpreter that early on makes it historically relevant.

Another well-known stack machine-based language is PostScript, a language used to describe documents for printing purposes. PostScript was developed at Xerox PARC in the late 1970s by John Warnock and others; it was based on an earlier language designed by John Warnock. The group eventually left PARC to start Adobe Systems.

2.5   FURTHER READING

Koopman, P. (1989). *Stack Computers: The New Wave.* Ellis Horwood Publisher. Available at
http://www.ece.cmu.edu/~koopman/stack_computers/
Synopsis: An introduction to stack machines, a not so new wave, but interesting nevertheless.

Rather, E., Colburn, D. and Moore, C. (1993). The evolution of Forth. *ACM SIGPLAN Notices* 28(3) – HOPL II, pp. 177–199.
Synopsis: Charles Moore is a maverick in the computing world, and everyone should know about his work. This paper tells the story of Forth.

Warnock, J. E. (2012). Simple ideas that changed printing and publishing. *Proceedings of the American Philosophical Society* 156(4): 363–378.
Synopsis: A historical perspective on PostScript, a stack machine language for printing.

2.6   GLOSSARY

Stack: A stack is "Last-In-First-Out" data structure. Its main operations are *push* – add an element to the top of the stack – and *pop* – remove the element from the top of the stack. Stacks play a critical role in programming language implementations well beyond Forth. Although usually invisible to the programmer, in virtually every modern programming language, a stack is the piece of memory that supports a thread of program execution. When procedures/functions are called, a block of data related to the parameters and return addresses is usually pushed onto the stack; subsequent procedure/function calls push other similar blocks onto the stack. Upon return, the corresponding blocks on the stack are usually popped.

Heap: The heap is another piece of memory underlying the implementation of many modern programming languages. The heap is used for dynamic memory allocation/deallocation, such as the creation of lists and objects. (Not to be confused with a data structure called Heap, a specialized tree-based data structure.)

Stack machine: A stack machine is a real or emulated computing machine that uses a stack, instead of registers, as its main support for evaluating the expressions of the program. Forth is a stack machine programming language. So are many modern virtual machines, such as the Java Virtual Machine.

2.7 EXERCISES

2.1 *Another language.* Implement the example program in another language, but preserve the style.

2.2 *Search.* In the example program, the comparison in line #78 is not in style, as it uses Python's high-level containment checking if x in y. Rewrite this piece of the program, i.e. searching whether a given word is in the dictionary on the heap, using go-forth style. Explain what happens to the performance of your program.

2.3 *True stack.* Python uses lists for implementing stacks, which makes the example program a bit confusing. Implement a true stack data structure in Python (possibly wrapping around a list) that provides the expected push, pop, peek and empty operations. Use your data structure in the example program, instead of the list introduced in line #7.

2.4 *A different task.* Write one of the tasks proposed in the Prologue using the go-forth style.

## CHAPTER 3

Arrays

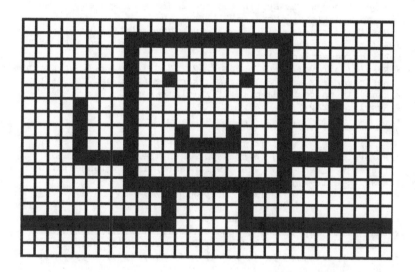

3.1 CONSTRAINTS

▷ Main data type: array – a fixed-size collection of elements.

▷ No explicit iteration; instead, an array is accessed by high-level, declarative operations.

▷ Computation unfolds as search, selection, and transformation of fixed-size data.

3.2   A PROGRAM IN THIS STYLE

```
 1  import sys, string
 2  import numpy as np
 3
 4  # Example input: "Hello  World!"
 5  characters = np.array([' ']+list(open(sys.argv[1]).read())+[' '])
 6  # Result: array([' ', 'H', 'e', 'l', 'l', 'o', ' ', ' ',
 7  #             'W', 'o', 'r', 'l', 'd', '!', ' '], dtype='<U1')
 8
 9  # Normalize
10  characters[~np.char.isalpha(characters)] = ' '
11  characters = np.char.lower(characters)
12  # Result: array([' ', 'h', 'e', 'l', 'l', 'o', ' ', ' ',
13  #             'w', 'o', 'r', 'l', 'd', ' ', ' '], dtype='<U1')
14
15  ### Split the words by finding the indices of spaces
16  sp = np.where(characters == ' ')
17  # Result: (array([ 0, 6, 7, 13, 14], dtype=int64),)
18  # A little trick: let's double each index, and then take pairs
19  sp2 = np.repeat(sp, 2)
20  # Result: array([ 0, 0, 6, 6, 7, 7, 13, 13, 14, 14], dtype=int64)
21  # Get the pairs as a 2D matrix, skip the first and the last
22  w_ranges = np.reshape(sp2[1:-1], (-1, 2))
23  # Result: array([[ 0,  6],
24  #               [ 6,  7],
25  #               [ 7, 13],
26  #               [13, 14]], dtype=int64)
27  # Remove the indexing to the spaces themselves
28  w_ranges = w_ranges[np.where(w_ranges[:, 1] - w_ranges[:, 0] > 2)]
29  # Result: array([[ 0,  6],
30  #               [ 7, 13]], dtype=int64)
31
32  # Voila! Words are in between spaces, given as pairs of indices
33  words = list(map(lambda r: characters[r[0]:r[1]], w_ranges))
34  # Result: [array([' ', 'h', 'e', 'l', 'l', 'o'], dtype='<U1'),
35  #          array([' ', 'w', 'o', 'r', 'l', 'd'], dtype='<U1')]
36  # Let's recode the characters as strings
37  swords = np.array(list(map(lambda w: ''.join(w).strip(), words)))
38  # Result: array(['hello', 'world'], dtype='<U5')
39
40  # Next, let's remove stop words
41  stop_words = np.array(list(set(open('../stop_words.txt').read().
       split(','))))
42  ns_words = swords[~np.isin(swords, stop_words)]
43
44  ### Finally, count the word occurrences
45  uniq, counts = np.unique(ns_words, axis=0, return_counts=True)
46  wf_sorted = sorted(zip(uniq, counts), key=lambda t: t[1], reverse=
       True)
47
48  for w, c in wf_sorted[:25]:
49      print(w, '-', c)
```

Note: If not familiar with Python, please refer to the Prologue (Pythonisms) for an explanation of lists, indexes and bounds.

3.3   COMMENTARY

THE MOST VISIBLE ELEMENT OF THIS STYLE is the concept of an array: a fixed-size collection of elements. All data is placed in arrays, and the sizes of these arrays are fixed, and must be known. Arrays can have one or more dimensions. A 1-dimensional array is called a *vector*, while an *N*-dimensional array is called an *N*-D *matrix*. When the data is smaller than the allotted slots in an array, the data is typically padded with some zero-like value through the end of the array.

Arrays, of course, are data structures that every programmer knows very well. But the mere use of arrays does not constitute the array programming style – far from it. A second, more important, constraint of this style is the absence of explicit iteration. Rather than explicitly iterating through the elements of arrays, as one might do in imperative languages, arrays are accessed with high-level, declarative operations that apply to the entire array at once. The high-level mathematical abstractions of array operations hide the low-level implementation details, and make these operations a perfect fit for highly parallel implementations, such as those supported by Graphical Processing Units (GPUs).

For example, consider the following snippet written in an imperative pseudo-language:

```
1 String[] cars = {'Volvo', 'BMW', 'Ford', 'Mazda'}
2 for (i = 0; i < cars.length; i++) {
3    cars[i] = cars[i].toLowerCase();
4 }
5 List<String> ocars = new List<String>();
6 for (int i = 0; i < cars.length; i++) {
7    if (cars[i].conatains('o'))
8       ocars.append(cars[i])
9 }
```

This snippet uses an array for placing the data (`cars`), but it is not written in array programming style. This other snippet, however, uses the array programming style:

```
1 String[] cars = {'Volvo', 'BMW', 'Ford', 'Mazda'}
2 cars = ToLowerCase(cars);
3 ocars = Where(cars.contains('o'))
```

Where the former snippet uses explicit iteration, the latter uses high-level, declarative operations on the array. Without these operations, we may be using arrays, but we are not using the array programming style.

In Python, collections of data are typically placed in variably sized lists, tuples, or dictionaries. Python also supports arrays, via the `array` module; however, those arrays are just the basic data structures, and do not support the array programming style. Python's lack of support for high-level array operations made some of its applications, especially in scientific computing, quite limited. Third-party libraries filled the void. The most popular of such libraries is **numpy**, a library that not only supports arrays but also supports powerful array operations. The example program uses numpy. Let's analyze it in detail.

At a high level, solving the term frequency problem in array style means placing all the textual data in an array, and then performing several array operations until we get the terms and their counts. In this implementation, we start with the raw data – an array of characters – in line #5. In order to simplify certain operations, white space is placed in the first and last positions of the array. Lines #10 and #11 show the first uses of the high-level array operations available in numpy. In them, the characters array is normalized by replacing all non-alphanumeric characters with white space and by transforming all characters to lowercase. The implementation of these high-level search-and-replace operations may perform several optimizations for parallel processing, but that is invisible to us.

Next, we need to tokenize, i.e. we need to identify the words in the array of characters. In order to stay true to the array programming style, this requires thinking about the problem differently than what we would normally think if we were using other kinds of data structures. The approach taken here is as follows: let's find the indices of the spaces; words, then, are sequences of characters between any two of those indices. We want to end up with a two-dimensional matrix where each row is a pair of [*start, end*] indices of words. To implement this approach, line #16 finds the indices of the spaces; then line #19 duplicates each of those numbers, in preparation for the two-dimensional matrix; the operation reshape in line #22 transforms the vector of duplicated indices into a 2D matrix; finally, line #28 selects only the rows where the difference between the *end* index and the *start* index is greater than 2, meaning that the word has at least two characters. At the end of that part of the program, in line #28, w_ranges contains the pairs [*start, end*] of all the words.

At this point in the program, we break from array programming style, and produce a variably sized list of words (line #33). This is because we cannot anticipate how many words there are, so we cannot use an array (unless we would assume some default maximum). The list words in line #33 is still a list of numpy character arrays. But from here on, we want to operate on words, not on characters. So in line #37 we create a new numpy array, this time with string elements (the words), rather than characters. Line #41 loads the stop words into an array of strings, and line #42 uses a powerful array operation for selecting only the words in the swords array that are not in the stop words array stop_words.

Finally, in line #45 another powerful array operation is used to return the unique words and their counts. Sorting, in line #46, is done in the traditional, non-array programming manner.

The example program is intermingled with comments showing a running input data example, so that the reader can better understand the meaning of the array operations. For that reason, the program seems longer than it really is. Without comments, the example program is quite concise:

```python
import sys, string
import numpy as np

# Split characters into words
characters = np.array([' ']+list(open(sys.argv[1]).read())+[' '])
characters[~np.char.isalpha(characters)] = ' '
characters = np.char.lower(characters)
sp = np.where(characters == ' ')
sp2 = np.repeat(sp, 2)
w_ranges = np.reshape(sp2[1:-1], (-1, 2))
w_ranges = w_ranges[np.where(w_ranges[:, 1] - w_ranges[:, 0] > 2)]
words = list(map(lambda r: characters[r[0]:r[1]], w_ranges))
swords = np.array(list(map(lambda w: ''.join(w).strip(), words)))

# Next, let's remove stop words
stop_words = np.array(list(set(open('../stop_words.txt').read().
    split(',')))) 
ns_words = swords[~np.isin(swords, stop_words)]

# Finally, count the word occurrences
uniq, counts = np.unique(ns_words, axis=0, return_counts=True)
wf_sorted = sorted(zip(uniq, counts), key=lambda t: t[1], reverse=
    True)

for w, c in wf_sorted[:25]:
    print(w, '-', c)
```

Typically, data-intensive programs written in array programming style tend to be small, concise, and, once we are familiar with the array operations, easy to read.

3.4 THIS STYLE IN SYSTEMS DESIGN

Array programming is inherently a set of intent expression ideas that come from mathematics. Nevertheless, some of these ideas have spilled over beyond mathematical applications. For starters, the powerful array operations for selecting, searching, and updating elements have found their way to the query languages of relational databases. Moreover, array programming, thought, for a time, to be a niche for engineering applications, is coming back with a vengeance in modern machine learning frameworks such as TensorFlow. Part X of this book covers these new developments.

3.5 HISTORICAL NOTES

Array programming is one of the oldest ideas in high-level programming. Computers were first developed to make calculations for scientific and engineering purposes. In science and engineering, linear algebra dominates. For that reason, the concept of multidimensional matrices, and operations on them, was, perhaps, one of the first concepts that made people want to elevate computer languages to notations that would be closer to mathematics than to assembly. The concepts of this programming style were first documented in the 1962 book *A Programming Language* by Kenneth Iverson, designer of APL. APL itself was an important and influential array programming language implemented at IBM in the 1960s, albeit a victim of its own obscure symbols. Its conciseness for describing array operations was unmatched at the time. "APL one-liners" – the ability to write a complex program in 1 line of APL code – became, at some point, a popular hobby among APL programmers.

Dartmouth BASIC adopted simple matrix operations in the mid-1960s. In the 1970s, the statistical programming system S (precursor of R) built on the same ideas. In the 1980s, MATLAB® came along as a modern programming environment for scientific and engineering applications. Like APL, MATLAB supports the powerful array operations that are characteristics of the array programming style. More recently, in the early 2000s, numpy was added to the Python ecosystem. Julia is an example of a modern language that supports vectorized operations.

3.6 FURTHER READING

Iverson, K. (1962). *A Programming Language*. Wiley. Available at
http://www.softwarepreservation.org/projects/apl
Synopsis: The rationale for, and detailed description of, APL.

3.7 GLOSSARY

Array: A fixed-size collection of data. Typically, the elements of an array are all of the same type, but that is not necessary. APL, for example, supported arrays where elements could be of different types.

Matrix: A multidimensional array.

Shape: The dimensions of an array. For example, a 3×2 matrix has shape (3, 2).

Vector: A 1-dimensional array.

Vectorization: Abstraction of iteration on elements of arrays as operations on the entire array.

3.8 EXERCISES

3.1 *Another language.* Implement the example program in another array programming language, such as MATLAB or Julia.

3.2 *Stay in style.* In line #33, followed by line #37, we broke from the array programming style to construct a list of words from the array of characters, and then create an array of strings from those words. Reimplement the second part of the program (from line #28 onward) without breaking from the array programming style. That is, work on an array of words expressed as a 2D array of characters.

3.3 *A different task.* Write one of the tasks proposed in the Prologue using the array programming style.

II

Basic Styles

This part of the book presents four basic styles: *Monolithic, Cookbook, Pipeline* and *Code Golf.* These are basic in the sense that their presence is pervasive in programming. All other styles use elements from these four.

## CHAPTER 4

Monolithic

4.1 CONSTRAINTS

▷ No named abstractions.

▷ No, or little, use of libraries.

4.2 A PROGRAM IN THIS STYLE

```python
1  #!/usr/bin/env python
2  import sys, string
3
4  # the global list of [word, frequency] pairs
5  word_freqs = []
6  # the list of stop words
7  with open('../stop_words.txt') as f:
8      stop_words = f.read().split(',')
9  stop_words.extend(list(string.ascii_lowercase))
10
11 # iterate through the file one line at a time
12 for line in open(sys.argv[1]):
13     start_char = None
14     i = 0
15     for c in line:
16         if start_char == None:
17             if c.isalnum():
18                 # We found the start of a word
19                 start_char = i
20         else:
21             if not c.isalnum():
22                 # We found the end of a word. Process it
23                 found = False
24                 word = line[start_char:i].lower()
25                 # Ignore stop words
26                 if word not in stop_words:
27                     pair_index = 0
28                     # Let's see if it already exists
29                     for pair in word_freqs:
30                         if word == pair[0]:
31                             pair[1] += 1
32                             found = True
33                             break
34                         pair_index += 1
35                     if not found:
36                         word_freqs.append([word, 1])
37                     elif len(word_freqs) > 1:
38                         # We may need to reorder
39                         for n in reversed(range(pair_index)):
40                             if word_freqs[pair_index][1] > \
                                    word_freqs[n][1]:
41                                 # swap
42                                 word_freqs[n], word_freqs[
                                        pair_index] = word_freqs[
                                        pair_index], word_freqs[n]
43                                 pair_index = n
44                 # Let's reset
45                 start_char = None
46         i += 1
47
48 for tf in word_freqs[0:25]:
49     print(tf[0], '-', tf[1])
```

4.3 COMMENTARY

IN THIS STYLE, even though we may be using a modern high-level programming language with powerful library functions available, the problem is solved in almost Good Old Times style: one piece of code, from beginning to end, with no new abstractions provided and not much use of the ones available in the libraries either. From a design perspective, the main concern is to obtain the desired output without having to think much about subdividing the problem or how to take advantage of code that already exists. Given that the entire problem is one single conceptual unit, the programming task consists of defining the data and control flow that rule this unit.

The example program works as follows. It holds a global list variable word_freqs (line #5) that is used for holding the pairs associating words with corresponding frequencies. The program first reads a list of stop words from a file, and extends with words comprising single letters, e.g. *a* (lines #7–9). Then the program engages in one long loop (lines #12–46) that iterates through the input file's lines, one by one. Inside that loop, there is a second, nested loop (lines #15–45) that iterates through each character of each line. The problem to be solved in this nested loop is to detect the beginning (lines #17–19) and the end (lines #21–43) of words, incrementing the frequency count for each detected word that is not a stop word (lines #26–43). In the case of a word that hasn't been seen before, a new pair is added to the word_freqs list, with count 1 (lines #35–36); in the case of a word that has already occurred (lines #29–34), the count is simply incremented (line #31). Since we want to print out the word frequencies by decreasing order of frequency, this program further ensures that the word_freqs list is always ordered by decreasing order of frequency (lines #39–43). At the end (lines #48–49), the program simply outputs the first 25 entries of the word_freqs list. Note that the only imports from the Python standard library are sys and string (line #3).

In the early days of computer programming, with low-level programming languages and relatively small programs, this style was all there was. Constructs such as goto gave further expression capabilities regarding control flow and therefore encouraged the existence of long "spaghetti code." Such constructs have been considered harmful for the development of all but the simplest programs, and are, for the most part, absent from modern programming languages. But goto statements are not the root cause of monolithic programs; when used responsibly, goto statements can result in nicely written programs. What is potentially bad from a maintenance point of view is the existence of long blocks of program text that fail to give the reader appropriate higher-level abstractions for what is going on.

It is possible to write programs in the monolithic style in any programming language, like the example program shows. In fact, it is not unusual to see long blocks of program text in modern programs. In some cases, those long program texts are warranted, for performance reasons or because the task being coded can't be easily subdivided. In many cases, however, they may be a symptom that the programmer didn't take the time to think carefully about the computational task at hand. Monolithic programs are often hard to follow, in the same way that a manual without chapters or sections would be hard to follow.

Cyclomatic complexity is a metric developed to measure the complexity of program texts, specifically the amount of control flow paths. As a rule of thumb, the higher the cyclomatic complexity of a piece of program text, the higher the chances that it is hard to understand. The computation of cyclomatic complexity sees program texts as directed graphs, and is given by the following formula:

$$CC = E - N + 2P \tag{4.1}$$

where

E = number of edges

N = number of nodes

P = number of exit nodes

For example, consider the following piece of program text:

```
1  x = raw_input()
2  if (x > 0):
3      print ''Positive''
4  else:
5      print ''Negative''
```

The directed graph of this text is as follows (node numbers correspond to line numbers):

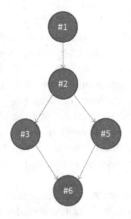

The cyclomatic complexity of this program text is
$CC = E - N + 2P = 5 - 5 + 2 = 2.$

The cyclomatic complexity has the same intention as metrics pertaining to measuring the readability of natural language texts such as the Flesch Reading Ease test and the Flesch-Kincaid grade level test. These metrics try to summarize style down to a number, and are backed up by some evidence in psychology regarding the difficulty that writing style has in people's understanding of texts. Clearly, writing programs is not the same as writing literary works. But when it comes to understanding what the writing is all about, there are many similarities. In some cases, a long piece of program text may be needed to make clear to the reader the inherent complexity of the programming task. More often, though, it probably isn't.

4.4   THIS STYLE IN SYSTEMS DESIGN

At the systems scale, monoliths are reflected in having single, large components that do everything that the application needs to do. This is in contrast to breaking down the system into modular subcomponents, each one responsible for a specific piece of functionality.

This style is considered bad practice at all scales. However, it is quite common to see monolithic code. It is important to recognize monoliths, and to try to understand the reasons that led to them.

4.5   FURTHER READING

Dijkstra, E. (1968). Go To statement considered harmful. *Communications of the ACM* 11(3): 147–148.
   Synopsis: Dijkstra rages against GOTO. A classic.

Knuth, D. (1974). Structured programming with go to statements. *ACM Computing Surveys* 6(4): 265–301.
   Synopsis: The best of many rebuttals of Dijkstra's rage against GOTO.

McCabe, T. (1976). A complexity measure. *IEEE Transactions on Software Engineering* SE-2(4): 308–320.
   Synopsis: Complexity metric for FORTRAN programs based on graphs. The first attempt at quantifying the cognitive load of various program design decisions.

4.6   GLOSSARY

Control flow: The order in which program statements are executed and program expressions are evaluated. Includes conditionals, iterations, function calls, returns, etc.

Cyclomatic complexity: A software metric that measures the number of linearly independent execution paths through a program's source code.

4.7 EXERCISES

4.1 *Another language.* Implement the example program in another language, but preserve the style.

4.2 *Readlines.* The example program reads one line at a time from the file. Modify it so that it first loads the entire file into memory (with readlines()), and then iterates over the lines in memory. Is this better or worse practice? Why?

4.3 *Two loops.* In lines #36–42 the example program potentially reorders, at every detected word, the list of word frequencies, so that it is always ordered by decreasing value of frequency. Within the monolithic style, modify the program so that the reordering is done in a separate loop at the end, before the word frequencies are printed out on the screen. What are the pros and cons of doing that?

4.4 *Cyclomatic complexity.* What is the cyclomatic complexity of the example program?

4.5 *A different task.* Write one of the tasks proposed in the Prologue using the monolithic style.

## CHAPTER 5

Cookbook

5.1 CONSTRAINTS

▷ No long jumps.

▷ Complexity of control flow tamed by dividing the large problem into smaller units using procedural abstraction. Procedures are pieces of functionality that may take input, but that don't necessarily produce output that is relevant for the problem.

▷ Procedures may share state in the form of global variables.

▷ The larger problem is solved by applying the procedures, one after the other, that change, or add to, the shared state.

5.2 A PROGRAM IN THIS STYLE

```python
#!/usr/bin/env python
import sys, string

# The shared mutable data
data = []
words = []
word_freqs = []

#
# The procedures
#
def read_file(path_to_file):
    """
    Takes a path to a file and assigns the entire
    contents of the file to the global variable data
    """
    global data
    with open(path_to_file) as f:
        data = data + list(f.read())

def filter_chars_and_normalize():
    """
    Replaces all nonalphanumeric chars in data with white space
    """
    global data
    for i in range(len(data)):
        if not data[i].isalnum():
            data[i] = ' '
        else:
            data[i] = data[i].lower()

def scan():
    """
    Scans data for words, filling the global variable words
    """
    global data
    global words
    data_str = ''.join(data)
    words = words + data_str.split()

def remove_stop_words():
    global words
    with open('../stop_words.txt') as f:
        stop_words = f.read().split(',')
    # add single-letter words
    stop_words.extend(list(string.ascii_lowercase))
    indexes = []
    for i in range(len(words)):
        if words[i] in stop_words:
            indexes.append(i)
    for i in reversed(indexes):
        words.pop(i)

def frequencies():
```

```
55      """
56      Creates a list of pairs associating
57      words with frequencies
58      """
59      global words
60      global word_freqs
61      for w in words:
62          keys = [wd[0] for wd in word_freqs]
63          if w in keys:
64              word_freqs[keys.index(w)][1] += 1
65          else:
66              word_freqs.append([w, 1])
67
68  def sort():
69      """
70      Sorts word_freqs by frequency
71      """
72      global word_freqs
73      word_freqs.sort(key=lambda x: x[1], reverse=True)
74
75
76  #
77  # The main function
78  #
79  read_file(sys.argv[1])
80  filter_chars_and_normalize()
81  scan()
82  remove_stop_words()
83  frequencies()
84  sort()
85
86  for tf in word_freqs[0:25]:
87      print(tf[0], '-', tf[1])
```

5.3 COMMENTARY

IN THIS STYLE, the larger problem is subdivided into subunits, *aka* procedures, each doing one thing. It is common in this style for the procedures to share data among themselves, as a means to achieve the final goal. Furthermore, the state changes may depend on previous values of the variables. The procedures are said to have *side effects* on this data. The computation proceeds with one procedure processing some data in the pool and preparing data for the next procedure. Procedures don't return data, as such, they just act on the shared data.

The example program is implemented as follows. A pool of shared data is declared (lines #5–7): the first variable, data, holds the contents of the input file; the second variable, words, holds the words that are extracted from the data; finally, the third variable, word_freqs, holds the word-frequency pairs. The three variables are all initialized to the empty list. This data is shared by a collection of procedures (lines #12–75), each doing a specific task:

- read_file(path_to_file) (lines #12–19) takes a path to a file and joins the entire contents of that file with the current value of the global variable data.

- filter_chars_and_normalize() (lines #21–30) replaces all non-alphanumeric characters in data with white space. The replacement is done in place.

- scan() scans the data (lines #32–39) for words by using the built-in function split, and it adds them to the global variable words.

- remove_stop_words() (lines #41–52) first loads the list of stop words from a file and appends it with single-letter words (lines #44–48). Then it traverses the words list and removes all the stop words from it. It does so by first storing the indexes of where the stop words are in the words list, and then by using the built-in function pop to remove those words from the words list.

- frequencies() (lines #54–66) traverses the words list and creates a list of pairs associating words with frequencies.

- sort() (lines #68–73) sorts the contents of the variable word_freqs by decreasing order of frequency. It does so by using the built-in function sort, which can take an anonymous function of two arguments, which, in this case, is the second element (index 1) of each pair in the word_freqs list.

The main program is from line #79 until the end. This piece of program text is where the cookbook nature of this style is most visible. Given that the larger problem is neatly divided into smaller subproblems, each addressed by a named procedure, the main program consists of issuing a sequence of commands corresponding to each of those procedures, similar to what one does when following an elaborate cooking recipe. In turn, each of those procedures changes the state of the shared variables, just like our following a cooking recipe changes the state of the ingredients.

A consequence of changing state over time (i.e. mutable state) is that procedures may not be *idempotent*. That is, calling a procedure twice may result in completely different states of the world, and completely different outputs for the program. For example, if we call the procedure read_file(path_to_file) twice, we end up with duplicate data in the data variable because of the cumulative nature of the assignment in line #19. An idempotent function or procedure is one that can be called multiple times yielding exactly the same observable effects as calling it just once. The lack of idempotency is seen by many as a source of programming errors.

5.4   THIS STYLE IN SYSTEMS DESIGN

In programming, this style is well suited for computational tasks that accumulate external data over time and whose behavior depends on that data. For example, for user interactions where the user is prompted for pieces of input at different points in time, maybe changing that input later, and where the outputs of the program depend on all data that the user has entered, holding on to state and changing it over time is a natural fit.

One issue that will be visible in later chapters is the granularity with which state is shared. In the example program, the variables are global, and they are shared by the entire collection of procedures. Global variables have long been considered a bad idea in all but the shortest programs. Many other styles discussed in this book use procedures that share variables in much smaller scopes. In fact, a lot of interesting stylistic work has been done over the years in order to limit side effects in specific manners.

At the systems scale, cookbook architectural styles are widely used in practice, with components sharing and changing external state such as that stored in databases.

5.5   HISTORICAL NOTES

During the 1960s, more and larger programs were being developed that challenged the programming technologies of the time. One of the main challenges had to do with programs being understood by other people. Programming languages were becoming increasingly more featureful, without them necessarily letting go of older constructs. The same program could be expressed in many different ways. In the late 1960s, a debate started about which features of the programming languages were "good" and which ones were "bad" [for the purpose of program understanding]. This debate, led, in part, by Dijkstra, advocated restraint in using some features considered harmful, such as long jumps (GOTO), and instead called for the use of higher-level iterative constructs (e.g. while loops), procedures and appropriate modularization of code. Not everyone agreed with Dijkstra's position, but his arguments prevailed. This gave rise to the era of *structured programming* – the style that is illustrated in this chapter, and that emerged in opposition to *unstructured*, or monolithic, programming as seen in the previous chapter.

5.6 FURTHER READING

Dijkstra, E. (1970). *Notes on Structured Programming*. Available from
http://www.cs.utexas.edu/users/EWD/ewd02xx/EWD249.PDF
Synopsis: Dijkstra was one of the most vocal advocates of structured programming. These notes lay out some of Dijkstra's thoughts on programming in general. A classic.

Wulf, W. and Shaw, M. (1973). Global variable considered harmful. *SIGPLAN Notices* 8(2): 28–34.
Synopsis: More opinions on structuring programs. This paper, as the title says, argues against global variables: not just structured programming, but *well*-structured programming.

5.7 GLOSSARY

Idempotence: A function, or procedure, is idempotent when multiple applications of it produce exactly the same observable effects as applying it just once.

Mutable variable: A mutable variable is one in which its assigned value can change over time.

Procedure: A procedure is a subroutine of a program. It may or may not receive input parameters, and it may or may not return a value.

Side effect: Side effect is a change in an observable part of a program. Side effects include writing to files or the screen, reading input, changing the value of observable variables, raising an exception, etc. Programs interact with the outside world via side effects.

5.8 EXERCISES

5.1 *Another language.* Implement the example program in another language, but preserve the style.

5.2 *Scope 'em.* Modify the example program so that there are no global variables but the imperative style is still the dominant style and the procedures are essentially the same.

5.3 *Double trouble.* In the example program, which procedures are idempotent and which ones aren't?

5.4 *Details matter.* Modify the example program as little as possible but make it so that all procedures become idempotent.

5.5 *A different program.* Write a different program, also in the Cookbook style, that does exactly the same thing as the example program but with different procedures.

5.6 *A different task.* Write one of the tasks proposed in the Prologue using the Cookbook style.

## CHAPTER 6

Pipeline

6.1 CONSTRAINTS

▷ Larger problem is decomposed using functional abstraction. Functions take input and produce output.

▷ No shared state between functions.

▷ The larger problem is solved by composing functions one after the other, in a pipeline, as a faithful reproduction of mathematical function composition $f \circ g$.

6.2   A PROGRAM IN THIS STYLE

```
1  #!/usr/bin/env python
2  import sys, re, operator, string
3
4  #
5  # The functions
6  #
7  def read_file(path_to_file):
8      """
9      Takes a path to a file and returns the entire
10     contents of the file as a string
11     """
12     with open(path_to_file) as f:
13         data = f.read()
14     return data
15
16 def filter_chars_and_normalize(str_data):
17     """
18     Takes a string and returns a copy with all nonalphanumeric
19     chars replaced by white space
20     """
21     pattern = re.compile('[\W_]+')
22     return pattern.sub(' ', str_data).lower()
23
24 def scan(str_data):
25     """
26     Takes a string and scans for words, returning
27     a list of words.
28     """
29     return str_data.split()
30
31 def remove_stop_words(word_list):
32     """
33     Takes a list of words and returns a copy with all stop
34     words removed
35     """
36     with open('../stop_words.txt') as f:
37         stop_words = f.read().split(',')
38     # add single-letter words
39     stop_words.extend(list(string.ascii_lowercase))
40     return [w for w in word_list if not w in stop_words]
41
42 def frequencies(word_list):
43     """
44     Takes a list of words and returns a dictionary associating
45     words with frequencies of occurrence
46     """
47     word_freqs = {}
48     for w in word_list:
49         if w in word_freqs:
50             word_freqs[w] += 1
51         else:
52             word_freqs[w] = 1
53     return word_freqs
54
```

```
55 def sort(word_freq):
56     """
57     Takes a dictionary of words and their frequencies
58     and returns a list of pairs where the entries are
59     sorted by frequency
60     """
61     return sorted(word_freq.items(), key=operator.itemgetter(1),
               reverse=True)
62
63 def print_all(word_freqs):
64     """
65     Takes a list of pairs where the entries are sorted by
             frequency and print them recursively.
66     """
67     if(len(word_freqs) > 0):
68         print(word_freqs[0][0], '-', word_freqs[0][1])
69         print_all(word_freqs[1:]);
70
71 #
72 # The main function
73 #
74 print_all(sort(frequencies(remove_stop_words(scan(
       filter_chars_and_normalize(read_file(sys.argv[1])))))))[0:25])
```

6.3 COMMENTARY

THE PIPELINE STYLE captures the model of a factory pipeline, where each station, or box, does one specific task over data that flows through it. In its purest form, the pipeline style is a faithful reflection of the theory of mathematical functions, where small boxes, *aka* functions, take input and produce output. In mathematics, a function is a relation that maps a set of inputs in one domain to a set of outputs in the same or another domain, where each input relates to exactly one output; for example $f(x) = x^2$ is a function that maps real numbers to non-negative real numbers such that when a value x is presented as input, the value x^2 is given as output. Like in a factory pipeline, functions can be combined with each other using mathematical function composition $f \circ g$ ("f after g"), as long as the output domain of the second function, g, is the same as, or is contained in, the input domain of the first, f. The input and output to functions can be anything, including other functions (boxes that do things themselves); functions that take functions as input or produce functions are called *higher-order functions*.

The pipeline programming style tries to achieve this kind of mathematical purity by seeing everything as relations mapping one set of inputs to one set of outputs. This constraint is very strong: in the pure pipeline style, the world outside boxed functions doesn't exist, other than in the beginning, as the source of input to a computation, and at the end, as the receiver of the output. The program needs to be expressed as boxed functions and function composition. Unfortunately, our term frequency program needs to read data from files, so it isn't completely pure. But it tries. In Chapter 25 we will see how to isolate impure actions from pure computation.

Let's analyze the example program. Similar to what has been done for the cookbook style, the problem of term frequency has been decomposed here into smaller problems, each one addressing a specific computational task. The decomposition is in all respects identical to the one used for the cookbook style example – the same procedures with the same names are used. But these procedures now have a special property: they each have one input parameter and return one value at the end. For example, read_file (lines #7–14) receives a string (the name of a file) as input and returns the contents of that file as output; filter_chars_and_normalize (lines #16–22) receives a string as input and returns a copy of that string with all non-alphanumeric characters replaced by whitespace and normalized to lowercase; etc. These procedures are now *functions* taking one input value and producing one output value. There is no state outside the functions.

Contrast this with the cookbook style program, where the procedures take no input, return nothing and simply make changes to the shared state. Note also the *idempotence* of these boxes as opposed to the lack of it in the procedures of the cookbook style. Idempotence means that calling each of these functions more than once yields exactly the same observable effects as calling them just once: the functions have no side effects in the observable world, and always produce the same output for a given input. For example, calling scan with the input ``hello world`` produces ['hello', 'world'] no matter how many times, and when, it is called.

The main program (from line #66 onward) is also indicative of the style: instead of a sequence of steps, we now have a chain of boxed functions, with the output of one function serving directly as input to the next. Mathematical function composition is written from right to left, in the sense that the function *after* is textually to the left of the previous function ($f \circ g =$ "f after g"), so reading programs in this style may feel a bit awkward for those not used to mathematics or to right-to-left languages. In this case, reading it sounds like "sort after frequencies after remove_stop_words... etc."; or, from right to left, "read_file then filter_chars_and_normalize... etc."

In the example program, all functions have one single argument, but, in general, functions may have multiple arguments. However, every multiple-argument function can be transformed in a sequence of single-value higher-order functions using a technique called *currying*. Consider, for example, the following function of three arguments:

```
def f(x, y, z):
    return x * y + z
```

which can then be called as such:

```
>>> f(2, 3, 4)
10
```

This function can be transformed in the following higher-order function:

```
def f(x):
    def g(y):
      def h(z):
          return x * y + z
      return h
    return g
```

which can then be called as such:

```
>>> f(2)(3)(4)
10
```

6.4 THIS STYLE IN SYSTEMS DESIGN

While it is hard to find systems with components that don't hold on to, and change, state in some form, the influence of pipelines can be found pervasively in computer systems engineering. One of the oldest and most well-known applications of this idea is the Unix shell *pipes*, where arbitrary commands can be sequenced together by binding the output of one with the input of the next using the character "|" (e.g. ps -ax | grep http). Each command in a piped chain is an independent unit that consumes input and produces output. Our term frequency program can be expressed nicely as a piped sequence of commands in any Linux shell:

```
grep -o ``[A-Za-z][A-Za-z][A-Za-z]*'' $1 \
  | tr '[:upper:]' '[:lower:]' \
  | grep -Ev ``^($(sed  -e 's/,/|/g' ../stop_words.txt))$'' \
  | sort | uniq -c | sort -rn | head -25 \
  | sed -e 's/^ *\([0-9]*\) *\([a-z]*\)/\2   -   \1/'
```

($1 on the first line stands for an argument to this shell script, intended to be the name of the file.)

The well-known MapReduce framework for data-intensive applications also embodies the constraints of the Pipeline style. We will cover it in more detail in Chapter 31.

The Pipeline style is particularly well suited for problems that can be modeled as pipelines in nature. Artificial intelligence algorithms such as graph searches, A*, etc. fall in this category. Compilers and other language processors are also a good fit, as they tend to consist of functions over graph and tree structures.

Besides problem fitness, there are good software engineering reasons for using this style, namely unit testing and concurrency. A program in this style is very easy to unit test, because it doesn't hold on to any state outside the testable boxed functions; running the tests once or multiple times or in different order always yields the same results. In imperative programming style, this invariance doesn't hold. For concurrency, again, the boxed functions are the units of computation, and they are independent of each other. Therefore it's straightforward to distribute them among multiple processors without any worries about synchronization and shared state. If a problem can be appropriately expressed in pipeline style, it's probably a good idea to do it!

6.5 HISTORICAL NOTES

In programming, functions are everywhere, even if this programming style isn't. Functions were invented multiple times by multiple people in multiple situations; the factory pipeline programming style is one that aims to stay faithful to mathematics, and hence it has thrived in only a niche of all the work related to functions.

The intense work in the theory of computation in the 1920s and 1930s that gave rise to Turing's work also had one mathematician, Alonzo Church, pursuing functions as the basis for computation. At about the same time that Turing published his work, Church showed how a very simple calculus with just three rules, the λ−calculus, could *transform* arbitrary inputs into outputs following specific relations. With this, he invented a universal symbol substitution machine that was as powerful as the Turing machine, but whose conceptual approach was quite different. A few years later, Turing's model of computation was accepted to be equivalent to Church's, in what's known as Kleene's Church-Turing thesis. But functional programming wouldn't exist, as such, until several decades later.

Functions were "invented" again, by necessity, in the course of the normal evolution of computers. They first appeared during the 1950s with the realization that in many programs some blocks of instructions needed to be executed many times during the execution of the program. This gave rise to the concept of the *subroutine*, and soon all languages supported that concept one way or another. Subroutines, in turn, found their way to higher-level programming languages either under that name or as *procedures* that could be called at any point of the program and that *return* to the caller context after completion. From procedures to functions was just a matter of supporting the existence of input parameters and output values. The second version of FORTRAN available in 1958, for example, had SUBROUTINE, FUNCTION, CALL and RETURN as language constructs.

But having functions in a programming language, and using them to write programs, is not the same as programming in pipelined functional style. As stated earlier on, the pipeline style is a strongly constrained style that aims at preserving the purity that exists in mathematical functions. Strictly speaking, a FORTRAN, C or Python "function" that affects the observable state of the world, in addition to implementing some relation between input and output, is not a function in the mathematical sense, and hence is considered outside the pipeline. Similarly, if a station in a factory pipeline counts the number of units that pass through it and stops the processing after some number has been processed, that is considered a side effect that breaks the purity of the pipeline model.

This programming style emerged during the 1960s in the context of LISP. LISP was designed to be a mathematical notation for computer programs, and was highly influenced by Church's λ−calculus. LISP was in sharp contrast with the dominant imperative programming languages of the time, precisely because of its strong functional style. LISP ended up departing from the λ−calculus purity, soon including constructs that allowed variables and mutable state. Nevertheless, its influence was substantial, especially in academic circles, where a new line of programming language design work emerged based on the functional style of programming introduced by LISP.

These days, the Pipeline, functional programming, style is supported by all main programming languages. As for the pure version of the Pipeline style, Haskell is leading the way.

6.6 FURTHER READING

Backus, J. (1978). Can programming be liberated from the von Neumann style? A functional style and its algebra of programs. *Communications of the ACM* 21(8): 613–641.

Synopsis: John Backus, of Backus-Naur Form (BNF) fame, radicalizes the discussion of programming languages by bashing the "complex, bulky, not useful" mainstream languages of the time and advocating pure functional programming. Despite its polarization, this paper touches on important problems in programming language design.

Church, A. (1936). An unsolvable problem of elementary number theory. *American Journal of Mathematics* 58(2): 345–363.
Synopsis: The original λ-calculus.

McCarthy, J. (1960). Recursive functions of symbolic expressions and their computation by machine, Part I. *Communications of the ACM* 3(4): 184–195.
Synopsis: Description of LISP and its relation to the λ-calculus.

Stratchey, C. (1967). Fundamental concepts in programming languages. Lecture notes. Reprinted in 2000 in *Higher-Order and Symbolic Computation* 13: 11–49, 2000.
Synopsis: Stratchey started the field of semantics of programming languages with his clear definitions for concepts and words that were being used carelessly and inconsistently, muddied by syntax. This paper is a write-up of lectures he gave in 1967. The differences between expressions and commands (statements), and functions and routines (procedures) are covered in this paper. Stratchey believed that side effects were important in programming and could have clear semantics. At the time of these lectures, he was involved in the definition of CPL, a research language that had a slow start and a faster disappearance, but was the grandmother of C.

6.7 GLOSSARY

Currying: Currying is a technique for transforming a function of multiple arguments into a sequence of higher-order functions, each with one single argument.

Function: In mathematics, a function is a relation that maps inputs to outputs. In programming, a function is a procedure that receives input and produces output. *Pure* functions, as in mathematics, don't have side effects. *Impure* functions are functions that have side effects.

Idempotence: A function, or procedure, is idempotent when multiple applications of it always produce the same observable effects as the first application.

Immutable variable: An immutable variable is one in which its assigned value never changes after the initial binding.

Side effects: A piece of code is said to have side effects when it modifies existing state or has an observable interaction with the world. Examples of side effects: modifying the value of a non-local variable or of an argument, reading/writing data from/to a file or the network or the display, raising an exception, and calling a function that has side effects.

6.8 EXERCISES

6.1 *Another language.* Implement the example program in another language, but preserve the style.

6.2 *2 in 1.* In the example program, the name of the file containing the list of stop words is hardcoded (line #36). Modify the program so that the name of the stop words file is given as the second argument in the command line. You must observe the following additional stylistic constraints: (1) no function can have more than 1 argument, and (2) the only function that can be changed is remove_stop_words; it's OK to change the call chain in the main block in order to reflect the changes in remove_stop_words.

6.3 *A different program.* Write a different program, also in the functional style, that does exactly the same thing as the example program, but with different functions.

6.4 *A different task.* Write one of the tasks proposed in the Prologue using the Pipeline style.

## CHAPTER 7

Code Golf

7.1 CONSTRAINTS

▷ As few lines of code as possible.

7.2 A PROGRAM IN THIS STYLE

```python
#!/usr/bin/env python
import re, sys, collections

stops = open('../stop_words.txt').read().split(',')
words = re.findall('[a-z]{2,}', open(sys.argv[1]).read().lower())
counts = collections.Counter(w for w in words if w not in stops)
for (w, c) in counts.most_common(25):
    print (w, '-', c)
```

7.3 COMMENTARY

THE MAIN CONCERN of this style is brevity. The goal is to implement the program's functionality in as few lines of code as possible. This is usually achieved by using advanced features of the programming language and its libraries. When brevity is the *only* goal, it is not unusual for this style to result in lines of code that are very long, with instruction sequences that are hard to understand. Often, too, textual brevity may result in programs that perform poorly or that have bugs, some of which only manifest themselves when the code is used in larger or different contexts. Brevity, however, when used appropriately, may result in programs that are quite elegant and easy to read because they are small.

The example program is possibly one of the shortest programs in terms of lines of code that can be written for implementing term frequency in Python.[1] Line #4 loads the list of stop words in just one line. It does so by chaining several file operations together: it opens the stop words file, reads the entire contents into memory, and then it splits the words around commas, obtaining a list of stop words bound to variable stops. Line #5 loads the list of words from the input file into memory in just one line. It does so by opening the file, reading its entire contents, and normalizing all characters to lowercase; after that it applies a regular expression to find all sequences of letters (a to z) with length greater than 2, to automatically eliminate single letters from the input file. The resulting list of words is placed in a list bound to the variable words. Line #6 uses Python's powerful collections library in order to obtain pairs of word-counts for all words that are not stop words. Finally, lines #7 and #8 print the 25 most frequent words and their counts. Line #7 uses, again, the powerful collections API, which provides a most_common method.

Brevity in terms of lines of code is related to the use of powerful abstractions that have already been created by someone else. Some languages' core libraries include a very large collection of utilities that come in handy for writing short programs; other languages' core libraries are small, and it is expected that utility libraries are provided by third parties. Python's built-in library is relatively large and varied. However, we could probably write an even shorter program by using a third-party library for natural language text processing (e.g. TextBlob). Indeed, if there's a utility program out there for computing term-frequency of a file, we could simply call one function like this: term_frequency(*file*, order='desc', limit=25).

While core libraries are usually trusted, when it comes to using third-party libraries, one needs to use some caution. By using an external library, we add a dependency between our code and someone else's project. It is not unusual for library developers to stop maintaining their code at some point, leaving the users of those libraries in limbo, especially when the source code is not

---

[1]This program is a slight improvement over the one contributed by Peter Norvig that is available in the GitHub repository for this book – see Preface.

available. Another issue is the stability, or lack thereof, of third-party code, which may introduce failures in our code.

7.4   THIS STYLE IN SYSTEMS DESIGN

One of the most popular metrics used by the software industry is Source Lines of Code (SLOC). For better or for worse, SLOC is used pervasively as a proxy for estimating cost, developer productivity, maintainability and many other management concerns. Many other metrics have come and gone over the years, but SLOC has survived them all. The Constructive Cost Model (COCOMO) is an example of a software cost estimation model based on SLOC. Developed in the 1970s, and updated a couple of times since then, this model is still widely used today.[2]

Clearly, at close inspection, SLOC is a poor estimate for some of those management concerns, especially programmer productivity, for which, unfortunately, SLOC is still used as a proxy (yes, there are still companies that assert that more SLOC/day = more programmer productivity!). The example program shown above is an extreme case: no one would say that the person who wrote the program in monolithic style is more productive than the person who wrote this beautiful small program. In general, correlations between SLOC and those higher-level management concerns such as cost and productivity have never been proven empirically; they are simply rough heuristics for making software project plans that may be useful in the beginning of a project.

On the popular culture front, brevity is seen as a sign of programming prowess, and the art of creating the shortest possible programs in various programming languages even has a name: *Code Golf*. However, trying to shorten programs for the sake of brevity alone is usually not a good idea. Oftentimes, the result is a small program that is very hard to read and that may also have some serious performance issues. Take, for example, the following term frequency program:

```python
1 #!/usr/bin/env python
2 import re, string, sys
3
4 stops = set(open("../stop_words.txt").read().split(",") + list(
        string.ascii_lowercase))
5 words = [x.lower() for x in re.split("[^a-zA-Z]+", open(sys.argv
        [1]).read()) if len(x) > 0 and x.lower() not in stops]
6 unique_words = list(set(words))
7 unique_words.sort(key=lambda x: words.count(x), reverse=True)
8 print("\n".join(["%s - %s" % (x, words.count(x)) for x in
        unique_words[:25]]))
```

This program has the exact same lines of code as the first one in this chapter. However, each of these lines is doing a lot more, and is expressed in a

---

[2]See, for example, http://ohloh.net's cost estimates for each project.

manner that is somewhat more difficult to understand. Let's look at line #5. This line is doing almost the same thing as line #5 of the first program: it's loading the words of the input file into memory, but filtering the stop words. This is done using a list comprehension, over a regular expression, after a couple of file system operations and a couple of tests! Lines #6 and #7 are even harder to understand: their goal is to produce a sorted list of unique words and their counts; first, line #6 computes the unique words by using the set data structure, which removes duplicates; then line #7 sorts those unique words using an anonymous function (a lambda in Python) that compares the counts of words pairwise.

This second program, though correct, performs very poorly. While the poor performance does not show in small text files, it will show quite dramatically in books like *Pride and Prejudice*. The reason for the poor performance is that the program keeps counting the words over and over again (line #7) whenever it needs those counts.

Even though brevity is usually a good goal that many programmers strive for, optimizing for LOC alone is a misguided goal, and may carry problems down the line that may be very difficult to diagnose.

7.5 HISTORICAL NOTES

Code golfs first emerged within APL (A Programming Language), a language developed in the 1960s by Ken Iverson, which included a large collection of non-standard symbols used as mathematical notation for manipulating arrays. By the early 1970s, a popular game had emerged among APL programmers consisting of coding useful functions in only one line (*aka* APL one-liners). Those one-liners tended to be relatively incomprehensible.

Code golfs can also be associated with the fewest keystrokes rather than with the fewest lines of code.

7.6 FURTHER READING

Boehm, B. (1981). *Software Engineering Economics*. Englewood Cliffs, NJ: Prentice-Hall.
> *Synopsis*: The dark art of software cost estimation, featuring the COCOMO model.

7.7 GLOSSARY

LOC: Lines of Code is one of the most widely used software metrics. Several variations of LOC exist. The most commonly used is Source LOC (SLOC), which counts only the lines with program instructions and ignores empty lines and lines with comments.

7.8 EXERCISES

7.1 *Another language.* Implement the example program in another language, but preserve the style.

7.2 *Fix it.* In the second example program, line #7 is a performance bottleneck. Fix it.

7.3 *Shorter.* Can you write a shorter term-frequency program in Python? If so, show it.

7.4 *A different task.* Write one of the tasks proposed in the Prologue using the code golf style.

III

Function Composition

.

This part of the book contains three styles related to function composition. *Infinite Mirror* shows the well-known mechanism of recursion, and it illustrates how to solve problems using its original concept, mathematical induction. *Kick Your Teammate Forward* is based on a programming approach known as continuation-passing style (CPS). *The One* is the first encounter in this book with a concept known as monad. The latter two styles use functions as regular data; this is one of the foundational offerings of functional programming.

## CHAPTER 8

Infinite Mirror

8.1 CONSTRAINTS

▷ All, or a significant part, of the problem is modeled using induction. That is, specify the base case (n_0) and then the $n + 1$ rule.

8.2 A PROGRAM IN THIS STYLE

```python
#!/usr/bin/env python
import re, sys, operator

# Mileage may vary. If this crashes, make it lower
RECURSION_LIMIT = 5000
# We add a few more, because, contrary to the name,
# this doesn't just rule recursion: it rules the
# depth of the call stack
sys.setrecursionlimit(RECURSION_LIMIT+10)

def count(word_list, stopwords, wordfreqs):
    # What to do with an empty list
    if word_list == []:
        return
    # The inductive case, what to do with a list of words
    else:
        # Process the head word
        word = word_list[0]
        if word not in stopwords:
            if word in wordfreqs:
                wordfreqs[word] += 1
            else:
                wordfreqs[word] = 1
        # Process the tail
        count(word_list[1:], stopwords, wordfreqs)

def wf_print(wordfreq):
    if wordfreq == []:
        return
    else:
        (w, c) = wordfreq[0]
        print(w, '-', c)
        wf_print(wordfreq[1:])

stop_words = set(open('../stop_words.txt').read().split(','))
words = re.findall('[a-z]{2,}', open(sys.argv[1]).read().lower())
word_freqs = {}
# Theoretically, we would just call count(words, stop_words,
#      word_freqs)
# Try doing that and see what happens.
for i in range(0, len(words), RECURSION_LIMIT):
    count(words[i:i+RECURSION_LIMIT], stop_words, word_freqs)

wf_print(sorted(word_freqs.items(), key=operator.itemgetter(1),
      reverse=True)[:25])
```

8.3 COMMENTARY

THIS STYLE encourages problem solving by induction. An inductive solution is one where a general goal is achieved in two steps: (1) solving one or more base cases, and (2) providing a solution that, if it works for the Nth case, also works for the $N^{th} + 1$ case. In computing, inductive solutions are usually expressed via *recursion*.

The example uses induction in two parts of the program: for counting the term frequencies (function count, lines #11–25) and for printing them out (function wf_print, lines #27–33). In both cases, the approach is the same. We check for the base case, the null list (lines #13–14 and #28–29), where recursion stops. Then we establish what to do for the general case; in the general case, we first process the head of the list (lines #18–23 and #31–32), followed by exercising the functions on the rest of the list (lines #25 and #33).

The example program includes an idiosyncratic element related to recursion in Python, expressed in lines #5–9 and then in line #40. As we recurse on the count function, the new calls take new portions of the stack; the stack is only popped at the end. But, existing on a finite amount of memory, the program eventually reaches a stack overflow. In order to avoid that, we first increase the recursion limit (line #9). But that is still not enough for a text as large as *Pride and Prejudice*. So instead of unleashing the count function on the entire list of words, we divide that list into N chunks, and call the function on one chunk at a time (lines #40–41). The function wf_print doesn't suffer from the same problem, because it only recurses 25 times.

In many programming languages, the problem of running into stack overflow on recursive calls is eliminated by a technique called *tail recursion optimization*. A *tail call* is a function call that happens as the very last action in a function. For example, in the following example, both calls to a and b are in tail positions of the function f:

```
1 def f(data):
2     if data == []:
3         a()
4     else:
5         b(data)
```

Tail recursion, then, is a recursive call that is in a tail position of the function. When that happens, language processors can safely eliminate the previous call's stack record altogether, as there is nothing else to do on that particular function call. This is called *tail recursion optimization*, and it effectively transforms recursive functions into loops, saving both space and time. Some programming languages (e.g. Haskell) do loops via recursion.

Unfortunately, Python doesn't do tail recursion optimizations, hence the idiosyncrasy of having to limit the depth of the call stack in the example.

8.4 HISTORICAL NOTES

Recursion has its origins in mathematical induction. The early programming languages of the 1950s, including Fortran, did not support recursive calls to subroutines. In the early 1960s some programming languages, starting with Algol 60 and Lisp, supported recursion, some of them with the use of explicit syntax. By the 1970s, recursion was commonplace in programming.

8.5 FURTHER READING

Daylight, E. (2011). Dijkstra's rallying cry for generalization: the advent of the recursive procedure, late 1950s–early 1960s. *The Computer Journal* 54(11). Available at
http://www.dijkstrascry.com/node/4
Synopsis: A retrospective look at the emergence of the idea of call stacks and recursion in programming.

Dijkstra, E. (1960). Recursive programming. *Numerische Mathematik* 2(1): 312–318.
Synopsis: Dijkstra's original paper describing the use of stacks for subroutine calls, as opposed to giving each subroutine its own memory space.

8.6 GLOSSARY

Stack overflow: Situation that happens when the program runs out of stack memory.

Tail recursion: A recursive call that happens as the very last action of a function.

8.7 EXERCISES

8.1 *Another language.* Implement the example program in another language, but preserve the style. Pay particular attention to whether the language you choose supports tail recursion optimization or not; if it does, your program should reflect that, rather than blindly copying the example Python program.

8.2 *More recursion.* Replace line #35 (loading and identification of stop words) with a function in infinite mirror style. What part of that line is easy to do in this style and what part is hard?

8.3 *No global counts.* The global variable word_freqs (line #37) is passed to count, which modifies its value. So the code relies on the order of side effects. Do this in the Pipeline style instead, by returning the word frequencies from count and passing the returned value to the recursive calls.

8.4 *A different task.* Write one of the tasks proposed in the Prologue using this style.

## CHAPTER 9

Kick Forward

9.1 CONSTRAINTS

Variation of the Pipeline style, with the following additional constraints:

▷ Each function takes an additional parameter, usually the last, which is another function.

▷ The function parameter is applied at the end of the current function.

▷ The function parameter is given, as input, what would be the output of the current function.

▷ The larger problem is solved as a pipeline of functions, but where the next function to be applied is given as a parameter to the current function.

9.2   A PROGRAM IN THIS STYLE

```python
1  #!/usr/bin/env python
2  import sys, re, operator, string
3
4  #
5  # The functions
6  #
7  def read_file(path_to_file, func):
8      with open(path_to_file) as f:
9          data = f.read()
10     func(data, normalize)
11
12 def filter_chars(str_data, func):
13     pattern = re.compile('[\W_]+')
14     func(pattern.sub(' ', str_data), scan)
15
16 def normalize(str_data, func):
17     func(str_data.lower(), remove_stop_words)
18
19 def scan(str_data, func):
20     func(str_data.split(), frequencies)
21
22 def remove_stop_words(word_list, func):
23     with open('../stop_words.txt') as f:
24         stop_words = f.read().split(',')
25     # add single-letter words
26     stop_words.extend(list(string.ascii_lowercase))
27     func([w for w in word_list if not w in stop_words], sort)
28
29 def frequencies(word_list, func):
30     wf = {}
31     for w in word_list:
32         if w in wf:
33             wf[w] += 1
34         else:
35             wf[w] = 1
36     func(wf, print_text)
37
38 def sort(wf, func):
39     func(sorted(wf.items(), key=operator.itemgetter(1), reverse=
          True), no_op)
40
41 def print_text(word_freqs, func):
42     for (w, c) in word_freqs[0:25]:
43         print(w, '-', c)
44     func(None)
45
46 def no_op(func):
47     return
48
49 #
50 # The main function
51 #
52 read_file(sys.argv[1], filter_chars)
```

9.3   COMMENTARY

I N THIS STYLE, functions take one additional parameter – a function – that is meant to be called at the very end, and passed what normally would be the return value of the current function. This makes it so that the functions don't return to their callers, and instead continue to some other function.

This style is known in some circles as *continuation-passing style*, and it is often used with anonymous functions (*aka* lambdas) as continuations, rather than with named functions. The example here uses named functions for readability.

The example program uses the same kind of decomposition that we have seen in previous styles, specifically the Pipeline style, so there is no need to repeat the explanation of the purpose of each function. What is noteworthy is the extra parameter seen in all of those functions, func, and how it is used: func is the next function after the current function finishes. Let's compare this program with the program written in the Pipeline style.

The main program in this style (line #52) calls one single function, read_file, giving it both the name of the file to read (coming from the command line input) and the next function to be called after read_file does its job – filter_chars. In contrast, in the Pipeline style, the main program is a chain of function calls that completely define all the steps of the "factory."

In this style, the read_file function (lines #7–10) reads the file and then, as its very last action, calls the func argument, which, in this case is the function filter_chars (as per line #52). As it does so, it gives it what would normally be its return value, data (see Pipeline style program, line #14) and next function to be called, normalize. The rest of the functions are designed in exactly the same manner. The chain of function calls is broken only when the no_op function is called in line #44; no_op does nothing.

9.4   THIS STYLE IN SYSTEMS DESIGN

This style can be used for different purposes. One of those purposes is compiler optimizations: some compilers transform the programs they compile into an intermediate representation that uses this style, so they can optimize for tail calls (see discussion in the previous chapter).

Another purpose is to deal with normal cases and failure cases: it may be convenient for a function to, in addition to the normal parameters, receive two functions as parameters that establish where to continue to if the function succeeds and if the function fails.

A third purpose is to deal with blocking Input/Output (IO) in single-threaded languages: in those languages, the programs never block until they reach an IO operation (e.g. waiting on network or disk), at which point the control is passed on to the next instruction of the program. Once the IO operation completes, the language runtime needs to continue where it left off, and that is done by sending one or more extra function arguments to functions. For example, the following code is part of the examples of Socket.io, a JavaScript library for WebSockets over Node.js:

```
1  function handler (req, res) {
2    fs.readFile(__dirname + '/index.html',
3    function (err, data) {
4      if (err) {
5        res.writeHead(500);
6        return res.end('Error loading index.html');
7      }
8
9      res.writeHead(200);
10     res.end(data);
11   });
12 }
```

In this example, the readFile function called in line 2 would, in principle, block the thread until the data is read from disk – that's the expected behavior in languages like C, Java, Lisp, Python, etc. In the absence of threads, that means that no requests would be served until the disk operation completes, which would be bad (disk accesses are slow). The design principle of JavaScript is to make asynchronicity a concern not of the application but of the underlying language processor. That is, the disk operation will block, but the application program continues to the next instruction – line #12 in this case, which is the return from the handler function. However, the language processor needs to be told what to do once the data is read, successfully or not, from the disk. That is achieved with a function as an extra parameter, an anonymous function defined in lines #3–11. Once the disk read operation unblocks and the main thread blocks on some other IO operation, the underlying language processor calls that anonymous function.

While used by necessity in JavaScript and other languages that don't support threads, when abused, this style can result in spaghetti code that is very hard to read (*aka* callback hell).

9.5 HISTORICAL NOTES

As is usually the case in all-things programming, this style has been "invented" by many people over the years, and for different purposes.

This style has its origins in GOTO, or jumps out of the blocks in the early 1960s. The earliest description of continuations, as such, dates back to 1964, in a presentation given by A. Wijngaarden, but, for a number of reasons, the idea did not catch on at the time. In the early 1970s, the idea emerged again in a few papers and presentations, as an alternative to GOTOs. From then on, the concept became well known within the programming language community. Continuations are used in the Scheme programming language, which first emerged in the late 1970s. These days, continuations are heavily used in logic programming languages.

9.6 FURTHER READING

Reynolds, J. (1993). The discoveries of continuations. *Lisp and Symbolic Computation* 6: 233–247.
  Synopsis: A retrospective look at the history of continuations.

9.7 GLOSSARY

Continuation: A continuation is a function representing "the rest of the program." This concept serves a variety of purposes, from optimizing compilers to providing denotational semantics to dealing with asynchronicity. It is also an alternative to language constructs such as goto statements and exceptions, as it provides a generalized mechanism for doing non-local returns from functions.

Callback hell: A form of spaghetti code that results from chaining anonymous functions as arguments several levels deep.

9.8 EXERCISES

9.1 *Another language.* Implement the example program in another language, but preserve the style.

9.2 *A different task.* Write one of the tasks proposed in the Prologue using this style.

## CHAPTER 10

The One

10.1 CONSTRAINTS

▷ Existence of an abstraction to which values can be converted.

▷ This abstraction provides operations to (1) wrap around values, so that they become the abstraction; (2) bind itself to functions, to establish sequences of functions; and (3) unwrap the value, to examine the final result.

▷ Larger problem is solved as a pipeline of functions bound together, with unwrapping happening at the end.

▷ Particularly for The One style, the bind operation simply calls the given function, giving it the value that it holds, and holds on to the returned value.

10.2   A PROGRAM IN THIS STYLE

```python
1  #!/usr/bin/env python
2  import sys, re, operator, string
3
4  #
5  # The One class for this example
6  #
7  class TFTheOne:
8      def __init__(self, v):
9          self._value = v
10
11     def bind(self, func):
12         self._value = func(self._value)
13         return self
14
15     def printme(self):
16         print(self._value)
17
18 #
19 # The functions
20 #
21 def read_file(path_to_file):
22     with open(path_to_file) as f:
23         data = f.read()
24     return data
25
26 def filter_chars(str_data):
27     pattern = re.compile('[\W_]+')
28     return pattern.sub(' ', str_data)
29
30 def normalize(str_data):
31     return str_data.lower()
32
33 def scan(str_data):
34     return str_data.split()
35
36 def remove_stop_words(word_list):
37     with open('../stop_words.txt') as f:
38         stop_words = f.read().split(',')
39     # add single-letter words
40     stop_words.extend(list(string.ascii_lowercase))
41     return [w for w in word_list if not w in stop_words]
42
43 def frequencies(word_list):
44     word_freqs = {}
45     for w in word_list:
46         if w in word_freqs:
47             word_freqs[w] += 1
48         else:
49             word_freqs[w] = 1
50     return word_freqs
51
52 def sort(word_freq):
53     return sorted(word_freq.items(), key=operator.itemgetter(1),
           reverse=True)
```

```
54
55 def top25_freqs(word_freqs):
56     top25 = ""
57     for tf in word_freqs[0:25]:
58         top25 += str(tf[0]) + ' - ' + str(tf[1]) + '\n'
59     return top25
60
61 #
62 # The main function
63 #
64 TFTheOne(sys.argv[1])\
65 .bind(read_file)\
66 .bind(filter_chars)\
67 .bind(normalize)\
68 .bind(scan)\
69 .bind(remove_stop_words)\
70 .bind(frequencies)\
71 .bind(sort)\
72 .bind(top25_freqs)\
73 .printme()
```

Note: If not familiar with Python, please refer to the Prologue (Pythonisms) for an explanation of the use of self and constructors in Python.

10.3 COMMENTARY

THIS STYLE is another variation in sequencing functions beyond the traditional function composition provided by most programming languages. In this style of composing functions, we establish an abstraction ("the one") that serves as the glue between values and functions. This abstraction provides two main operations: a *wrap* operation that takes a simple value and returns an instance of the glue abstraction, and a *bind* operation that feeds a wrapped value to a function.

This style comes from the Identity monad in Haskell, a functional programming language where functions are not allowed to have side effects of any kind. Because of this strong constraint, Haskell designers have come up with interesting approaches to things that most programmers take for granted – like state and exceptions – and they did it using an elegant uniform approach called *monad*.

Rather than explaining what monads are, let's analyze the style used in the example program. Lines #7–15 define the glue abstraction for this example, TFTheOne. Note that we are modeling it as a class, instead of a set of independent functions (more about this in the exercises below). TFTheOne provides a constructor and 2 methods, bind and printme. The constructor takes a value and makes the newly created instance hold on to that value (line #9); in other words, the constructor *wraps* a TFTheOne instance around a given value. The bind method takes a function, calls it giving it the value that the instance is holding on to, updates the internal value, and returns the same TFTheOne instance; in other words, the bind method feeds a value into a function and returns the instance that wraps the new result of that function application. Finally, the printme method prints the value onto the screen.

The functions defined in lines #21 through #59 are generally similar to the functions that we have seen in previous styles, specifically the Pipeline style, so there is no need to explain what they do.

The interesting part of this program is from line #64 onward, the main portion of the program. That block chains the functions together, from left to right, binding the return values with the next functions to be called in the sequence, using the TFTheOne abstraction as the glue for that chain. Note that for all practical purposes, and ignoring minor differences, this line plays the same role as line #66 in the Pipeline style:

```
1 word_freqs = sort(frequencies(remove_stop_words(scan(
        filter_chars_and_normalize(read_file(sys.argv[1]))))))
```

Most programming languages have come to provide the style of the line above as the norm of composing functions. However, like the Kick Forward style, The One style does the composition in its own unique manner. Unlike the Kick Forward style, however, The One style, by itself, doesn't have distinguishing properties that make it desirable to use in practice, except, perhaps, the interesting property of allowing us to write function chains from left to right instead of right to left. Indeed, the Identity monad is considered a trivial monad, because it doesn't do anything interesting with the functions that it handles – it just calls them. But not all monads are like that.

All monads have essentially the same interface as TFTheOne: a wrapper operation (i.e. the constructor), a bind operation, and some operation that shows what's inside the monad. But these operations can do different things, resulting in different monads – i.e. different ways of chaining computations. We will see another example in Chapter 25, Quarantine.

10.4 HISTORICAL NOTES

Monads have their origins in category theory. They were brought to programming languages in the early 1990s in the context of the Haskell programming language, in an effort to model the incorporation of side effects into pure functional languages.

10.5 FURTHER READING

Moggi, E. (1989). An abstract view of programming languages. Lecture Notes produced at Stanford University.
Synopsis: With these notes, Moggi brought category theory into the realm of programming languages.

Wadler, P. (1992). The essence of functional programming. *19th Symposium on Principles of Programming Languages*, ACM Press.
Synopsis: Wadler introduces monads in the context of pure functional programming languages.

10.6 GLOSSARY

Monad: A structure (for example, an object) that encapsulates computations defined as a sequence of steps. A monad has two main operations: (1) a constuctor that wraps a value within the monad, (2) a bind operation that takes a function as argument, binds it to the monad in some way, and returns a monad (maybe itself). Additionally, a third operation is used to unwrap/print/evaluate the monad.

10.7 EXERCISES

10.1 *Another language.* Implement the example program in another language, but preserve the style.

10.2 *Class vs. functions.* The Identity monad in the example is expressed as a class, TFTheOne. In functional programming languages, the monadic operations are simple wrap and bind functions. wrap takes a simple value and returns a function that, when called, returns the value; bind takes a wrapped value and a function, and returns the result of calling that function on the application of the wrapped value. Redo the example by defining these two functions and using them in the following manner:

```
printme(..wrap(bind(wrap(sys.argv[1]),read_file),filter_chars)..)
```

10.3 *A different task.* Write one of the tasks proposed in the Prologue using this style.

IV

Objects and Object Interaction

There are many ways of abstracting a problem, a concept, or an observable phenomenon. The *Monolithic* style is a baseline illustration of what it is like when the problem is not abstracted and, instead, is solved in all its concreteness and detail. The example shown in the *Code Golf* style also doesn't abstract the problem, as such; but because it uses powerful abstractions provided by the programming language and its libraries, almost every line in that program captures a conceptual unit of thought, even though those units don't have explicit names. The *Cookbook* style uses procedural abstraction: the larger problem is decomposed as a series of steps, or procedures, each with a name, that operate over a pool of shared data. The *Pipeline* style program uses functional abstraction: the larger problem is decomposed as a collection of functions, each with a name, that take input, produce output, and combine with each other by providing the output of one function as the input of another.

This part of the book contains a collection of styles related to the *Object* abstraction. Ask programmers what an object is, and likely you will see the answers converging to four or five main concepts, all related, but slightly different from each other. As part of that variety, we can identify a number of different mechanisms by which objects are thought to interact with each other. This collection of styles reflects that variety.

## CHAPTER 11

Things

11.1 CONSTRAINTS

▷ The larger problem is decomposed into *things* that make sense for the problem domain.

▷ Each *thing* is a capsule of data that exposes procedures to the rest of the world.

▷ Data is never accessed directly, only through these procedures.

▷ Capsules can reappropriate procedures defined in other capsules.

11.2 A PROGRAM IN THIS STYLE

```python
#!/usr/bin/env python
import sys, re, operator, string
from abc import ABCMeta

#
# The classes
#
class TFExercise():
    __metaclass__ = ABCMeta

    def info(self):
        return self.__class__.__name__

class DataStorageManager(TFExercise):
    """ Models the contents of the file """

    def __init__(self, path_to_file):
        with open(path_to_file) as f:
            self._data = f.read()
        pattern = re.compile('[\W_]+')
        self._data = pattern.sub(' ', self._data).lower()

    def words(self):
        """ Returns the list words in storage """
        return self._data.split()

    def info(self):
        return super(DataStorageManager, self).info() + ": My
            major data structure is a " + self._data.__class__.
            __name__

class StopWordManager(TFExercise):
    """ Models the stop word filter """

    def __init__(self):
        with open('../stop_words.txt') as f:
            self._stop_words = f.read().split(',')
        # add single-letter words
        self._stop_words.extend(list(string.ascii_lowercase))

    def is_stop_word(self, word):
        return word in self._stop_words

    def info(self):
        return super(StopWordManager, self).info() + ": My major
            data structure is a " + self._stop_words.__class__.
            __name__

class WordFrequencyManager(TFExercise):
    """ Keeps the word frequency data """

    def __init__(self):
        self._word_freqs = {}
```

```
51    def increment_count(self, word):
52        if word in self._word_freqs:
53            self._word_freqs[word] += 1
54        else:
55            self._word_freqs[word] = 1
56
57    def sorted(self):
58        return sorted(self._word_freqs.items(), key=operator.
              itemgetter(1), reverse=True)
59
60    def info(self):
61        return super(WordFrequencyManager, self).info() + ": My
              major data structure is a " + self._word_freqs.
              __class__.__name__
62
63 class WordFrequencyController(TFExercise):
64    def __init__(self, path_to_file):
65        self._storage_manager = DataStorageManager(path_to_file)
66        self._stop_word_manager = StopWordManager()
67        self._word_freq_manager = WordFrequencyManager()
68
69    def run(self):
70        for w in self._storage_manager.words():
71            if not self._stop_word_manager.is_stop_word(w):
72                self._word_freq_manager.increment_count(w)
73
74        word_freqs = self._word_freq_manager.sorted()
75        for (w, c) in word_freqs[0:25]:
76            print(w, '-', c)
77
78 #
79 # The main function
80 #
81 WordFrequencyController(sys.argv[1]).run()
```

Note: If not familiar with Python, please refer to the Prologue (Pythonisms) for an explanation of the use of self and constructors (__init__) in Python.

11.3 COMMENTARY

IN THIS STYLE, the problem is divided into collections of procedures, each collection sharing, and hiding, a main data structure and/or control. Those capsules of data and procedures are called *things* or *objects*. Data is never directly accessed from outside these *things*; instead it is hidden in them, and accessed only through the exposed procedures, also called *methods*. From the point of view of callers, *things* can be replaced by other *things* with different sets of implementations, as long as the interface to the procedures is the same.

This style of programming comes in many flavors, and we will see some of them in other chapters. Mainstream Object-Oriented Programming (OOP) languages such as Java, C# and C++ lump together several concepts that define what has come to be expected out of *objects*. The essence of the Things style is simply this: procedures that share data among themselves, hiding it from the outside world. This style can be achieved not just with an OOP language but also with any other language that supports imperative features. Additionally, this style is often associated with *classes* and *inheritance*, although strictly speaking, these concepts are neither necessary nor sufficient for programming in Things style.

The example program takes advantage of Python's OOP features, but the explanation that follows focuses, first, on the essentials of this style. In the example, the problem is modeled with four main objects: the WordFrequencyController, the DataStorageManager, the StopWordManager and the WordFrequencyManager:

- The DataStorageManager (lines #14–28) is concerned with getting the text data from the outside world and distilling it into words that the rest of the application can use. Its main public method, `words`, returns the list of words in storage. To that effect, its constructor first reads the file, cleans the data out of non-alphanumeric characters and normalizes it to lowercase. The DataStorageManager object *hides* the text data, providing only a procedure to retrieve the words in it. This is a typical example of *encapsulation* in object-oriented programming. The DataStorageManager is an abstraction of data and behavior related to the input text data of the problem at hand.

- StopWordManager (lines #30–43) offers a service to the rest of the application that consists of determining if a given word is a stop word or not. Internally, it holds a list of stop words against which it performs its service `is_stop_word`. Also here, we see encapsulation at work: the rest of the application doesn't need to know what kind of data the StopWordManager uses internally, and how it decides whether a given word is a stop word or not. This object exposes the procedure `is_stop_word`, which is the one that the rest of the application needs to know about.

- The WordFrequencyManager (lines #45-61) is concerned with managing the word counts. Internally, it revolves around one main data structure, a dictionary that maps words to counts. Externally, it provides procedures that can be called by the rest of the application, namely: `increment_count` and `sorted`. The first, `increment_count`, changes the state of the object by changing the internal dictionary; `sorted` returns a collection of the words sorted by their frequency.

- The WordFrequencyController (lines #63-76) is the object that starts it all. Its constructor instantiates all the other application objects, and its main method is simply called `run`. That method polls the DataStorageManager for words, tests if each of those words is a stop word by asking the StopWordManager and, if not, it invokes the WordFrequencyManager for it to increment the count for that word. After iterating through all the words provided by the DataStorageManager, the method `run` retrieves a sorted (by frequency count) collection of words from the WordFrequencyManager and displays the 25 most frequently occurring words.

- Finally, the main method simply instantiates a WordFrequencyController and invokes its `run` method.

As customary when using mainstream OOP languages, including Python, the example program uses *classes* to define the capsules of data and procedures. Classes (lines #14, #30, #45 and #63) are templates for the construction of objects. As mentioned before, even though the most popular OOP languages lead us to believe that classes are central to OOP, the fact is that they are neither necessary nor sufficient for programming in the Things style. They are simply a mechanism for defining objects, one that is particularly convenient when the applications use many objects (*aka instances*) that are of a similar kind. But many languages that support this style of programming, most notably JavaScript, don't support the concept of classes explicitly, so objects are defined by other means – e.g. functions, dictionaries, etc.

Inheritance is another concept that has made it to all mainstream OOP languages. The example program uses inheritance in a somewhat artificial manner, simply for illustration purposes. In the example, an *abstract base class* called TFExercise is defined (lines #8–12). The statement in line #9 is Python's way of saying that this class is *abstract*, meaning that it cannot be used directly to create objects and, instead, it is meant to be extended by other classes. The TFExercise class defines only one method, `info`, meant to print out information about the each class in the example. All classes in our example inherit from TFExercise – in Python, inheritance is established with the syntax `class A(B)` meaning that class A extends class B. Three of our four classes *override* the method `info` of the superclass – lines #27–28, #42–43 and #60–61. The WordFrequencyController class doesn't; instead, it (re-)uses the method defined in the superclass as if it were its own. (Note that none of these methods are being called in the example – they are used in the Exercises section.)

At its core, inheritance is a modeling tool, a central part of the conceptual tools for modeling the real world. Inheritance captures the *is a* relationship between objects, or classes of objects: for example, a car *is a* vehicle, therefore whatever general state and procedures vehicles have, should also be present in cars. In programming, inheritance is also a mechanism for (re-)using code. In the vehicle-car example, the procedures and internal implementation for vehicle objects can be made available to car objects, even though car objects may *extend* or *override* what vehicles do.

Inheritance is often associated with classes, as in the example program, but, again, both concepts are independent of each other. Inheritance can also be established between objects directly; OOP languages that don't support classes may still support inheritance between objects – such is the case with JavaScript. In its essence, inheritance in programming is the ability to use pre-existing object definitions as part of the definition of new objects.

The Things style is particularly well suited for modeling objects of the real world such as originally envisioned by Simula 67. Graphical User Interface (GUI) programming is a particularly well-suited domain for this style. The dominance of C++ and Java over the past couple of decades has made an entire generation of programmers think of modeling computational problems using this style, but it is not always the best fit for everything.

11.4   THIS STYLE IN SYSTEMS DESIGN

With the interest in OOP in the 1980s and 1990s being so great, many people tried to use the same principles in large-scale distributed systems. The approach is generally known as "distributed objects," and it gathered intense attention from industry during the 1990s – before, and at about the same time as, the advent of the Web. The main idea at systems design level is that software in one node of the Internet can acquire references to remote objects residing in another node; from then on, it can invoke the methods of those remote objects, almost as if they were local – the invocation looks identical to local object invocations. The platforms and frameworks supporting distributed objects automate, to a large extent, the generation of stubs and network code, making the lower level virtually invisible for application programmers.

Examples of such platforms and frameworks include CORBA (Common Object Request Broker Architecture, a standard established by a large committee) and Java's Remote Method Invocation.

While an interesting idea in principle, this approach has somewhat failed to gain traction in practice. One of the main problems of distributed objects is that the system needs to adopt a common programming language and/or a common infrastructure for all its distributed components. Distributed systems often need to access components developed by different groups of people at different points in time, so the assumption of a common infrastructure tends not to hold.

Additionally, the large standardization effort of CORBA collided with the emergence and exponential adoption of the Web, which is based on a different approach to large-scale systems design. Nevertheless, distributed objects is an interesting approach to systems design, one that is well aligned with the smaller scale programs written in the Things style.

11.5 HISTORICAL NOTES

Object-oriented programming was first introduced by Simula 67 in the 1960s. Simula 67 already had all the main concepts explained above, namely objects, classes and inheritance. Soon after, in the early 1970s, Xerox's investment in creating the personal computer included the effort to design and implement Smalltalk, a language that was very much inspired by Simula but designed for programming the "modern" (at the time) graphical displays and new peripherals. Smalltalk's style is more similar to the one in the next chapter, though, so it will be covered there.

11.6 FURTHER READING

Dahl, O.-J., Myhrhaug, B. and Nygaard, K. (1970). Common Base Language. Technical report, Norwegian Computing Center, available at
http://www.fh-jena.de/ kleine/history/languages/Simula-CommonBaseLanguage.pdf
Synopsis: The original description of Simula.

11.7 GLOSSARY

Abstract class: A class that is not meant to be directly instantiated and, instead, is used solely as a unit of behavior that can be inherited by other classes.

Base class: A class from which another class inherits. Same as superclass.

Class: A template of data and procedures for creating (*aka* instantiating) objects.

Derived class: A class that inherits from another class. Same as subclass.

Extension: The result of defining an object, or a class of objects, using another object, or class, as the base and adding additional data and procedures.

Inheritance: The ability to use pre-existing object, or class, definitions as part of the definition of new objects or classes.

Instance: A concrete object, usually one that has been constructed from a class.

Method: A procedure that is part of an object or class.

Object: A capsule of data and procedures.

Overriding: Changing an inherited procedure by giving it a new implementation in a subclass.

Singleton: An object that is the sole instance of a class of objects.

Subclass: Same as derived class.

Superclass: Same as base class.

11.8 EXERCISES

11.1 *Another language.* Implement the example program in another language, but preserve the style.

11.2 *Forgotten methods.* In the example program, the methods `info` are never invoked. Change the program (to the minimum) so that all of them are invoked, and their results are printed out. What happens internally when that method is invoked on the DataStorageManager object and on the WordFrequencyController object? Explain the results.

11.3 *A different program.* Write a different program, also in the Things style, that does exactly the same thing as the example program, but with different classes, i.e. using a different decomposition of the problem. No need to preserve the relatively artificial inheritance relations of the example, or the `info` methods.

11.4 *Classless.* Write a Things style implementation of the term frequency problem without using Python's classes. No need to preserve the `info` methods.

11.5 *A different task.* Write one of the tasks proposed in the Prologue using this style.

## CHAPTER 12

Letterbox

12.1 CONSTRAINTS

▷ The larger problem is decomposed into *things* that make sense for the problem domain.

▷ Each *thing* is a capsule of data that exposes one single procedure, namely the ability to receive and dispatch messages that are sent to it.

▷ Message dispatch can result in sending the message to another capsule.

12.2   A PROGRAM IN THIS STYLE

```python
#!/usr/bin/env python
import sys, re, operator, string

class DataStorageManager():
    """ Models the contents of the file """
    _data = ''

    def dispatch(self, message):
        if message[0] == 'init':
            return self._init(message[1])
        elif message[0] == 'words':
            return self._words()
        else:
            raise Exception("Message not understood " + message
                [0])

    def _init(self, path_to_file):
        with open(path_to_file) as f:
            self._data = f.read()
        pattern = re.compile('[\W_]+')
        self._data = pattern.sub(' ', self._data).lower()

    def _words(self):
        """ Returns the list words in storage"""
        data_str = ''.join(self._data)
        return data_str.split()

class StopWordManager():
    """ Models the stop word filter """
    _stop_words = []

    def dispatch(self, message):
        if message[0] == 'init':
            return self._init()
        elif message[0] == 'is_stop_word':
            return self._is_stop_word(message[1])
        else:
            raise Exception("Message not understood " + message
                [0])

    def _init(self):
        with open('../stop_words.txt') as f:
            self._stop_words = f.read().split(',')
        self._stop_words.extend(list(string.ascii_lowercase))

    def _is_stop_word(self, word):
        return word in self._stop_words

class WordFrequencyManager():
    """ Keeps the word frequency data """
    _word_freqs = {}

    def dispatch(self, message):
        if message[0] == 'increment_count':
```

```
53            return self._increment_count(message[1])
54        elif message[0] == 'sorted':
55            return self._sorted()
56        else:
57            raise Exception("Message not understood " + message
                  [0])
58
59    def _increment_count(self, word):
60        if word in self._word_freqs:
61            self._word_freqs[word] += 1
62        else:
63            self._word_freqs[word] = 1
64
65    def _sorted(self):
66        return sorted(self._word_freqs.items(), key=operator.
              itemgetter(1), reverse=True)
67
68 class WordFrequencyController():
69
70    def dispatch(self, message):
71        if message[0] == 'init':
72            return self._init(message[1])
73        elif message[0] == 'run':
74            return self._run()
75        else:
76            raise Exception("Message not understood " + message
                  [0])
77
78    def _init(self, path_to_file):
79        self._storage_manager = DataStorageManager()
80        self._stop_word_manager = StopWordManager()
81        self._word_freq_manager = WordFrequencyManager()
82        self._storage_manager.dispatch(['init', path_to_file])
83        self._stop_word_manager.dispatch(['init'])
84
85    def _run(self):
86        for w in self._storage_manager.dispatch(['words']):
87            if not self._stop_word_manager.dispatch(['is_stop_word
                  ', w]):
88                self._word_freq_manager.dispatch(['increment_count
                      ', w])
89
90        word_freqs = self._word_freq_manager.dispatch(['sorted'])
91        for (w, c) in word_freqs[0:25]:
92            print(w, '-', c)
93
94 #
95 # The main function
96 #
97 wfcontroller = WordFrequencyController()
98 wfcontroller.dispatch(['init', sys.argv[1]])
99 wfcontroller.dispatch(['run'])
```

12.3 COMMENTARY

THIS STYLE takes a different perspective on the *Things* concept explained in the previous chapter. The application is divided in exactly the same way. However, rather than the *Things* (*aka* objects) exposing a set of procedures to the outside world, they expose one single procedure: that of accepting messages. Data and procedures are hidden. Some messages are understood by the objects, and acted upon by means of execution of procedures; others are not understood, and are either ignored or produce some form of error; others may be processed not directly by the object but by other objects that have some relation to the receiving object.

In the example program, the solution uses essentially the same entities as in the previous example, but without exposing the methods. Instead, the classes, all of them, expose only one method, dispatch, that takes a message – see these methods in lines #8–14, #31–37, #51–57 and #70–76. The message consists of a tag that identifies it, and zero or more arguments that carry data for the internal procedure. Depending on the tag of the message, internal methods may be called, or an exception "Message not understood" is raised. Objects interact by sending messages to each other.

Styles based on object abstractions don't necessarily require inheritance, although these days most programming environments supporting Object-Orientated Programming (OOP) also support inheritance. An alternative reuse mechanism that achieves something very similar, but that is particularly fit for the message dispatch style, is *delegation*. The example program doesn't show it, but it is possible for the dispatch methods to send the message to another object when they don't have a method for it. The programming language Self, for example, gave objects *parent* slots that programmers could set dynamically; when an object receives a message to execute some action, if no such action is found within the object, the message is forwarded to its parent(s).

12.4   THIS STYLE IN SYSTEMS DESIGN

In distributed systems, and without further abstractions, components interact by sending messages to each other. The message-passing style of OOP is a much better fit than remote procedure/method call for distributed systems design: messages carry a much lower overhead in terms of interfacing components.

12.5   HISTORICAL NOTES

The *Letterbox* style illustrates the mechanism of message dispatch that underlies all OOP languages, at least conceptually. It is particularly similar to Smalltalk (1970s), historically one of the most important OOP languages. Smalltalk was designed under the principle of stylistic purity around objects;

rather than trying to amass a large number of useful programming features that had been seen in other languages, Smalltalk focused on trying to achieve conceptual consistency by treating everything as objects, their interaction being via messages – a purity goal similar to what certain functional languages do with the concept of functions. In Smalltalk, everything, including numbers, for example, is an object; classes are objects too; etc.

Variations of this style also appear in concurrent programming, specifically in the Actor model, which will be seen in a later chapter.

12.6 FURTHER READING

Kay, A. (1993). The Early History of Smalltalk. *HOPL-II,* ACM, New York, pp. 69–95.
    Synopsis: The history of Smalltalk told by Alan Kay, one of its creators.

12.7 GLOSSARY

Delegation: The ability of an object to use methods from another object when requested to execute a procedure.

Message dispatch: The process of receiving a message, parsing its tag, and determining the course of action, which can be a method execution, the return of an error or the forwarding of the message to other objects.

12.8 EXERCISES

12.1 *Another language.* Implement the example program in another language, but preserve the style.

12.2 *Delegation.* Let's bring back the `info` methods of the Things style (previous chapter). Write a version of the program without using Python's inheritance relations, but in a way that preserves those relations' code reuse intention. That is, the `info` methods should be available to all classes when they receive the `info` message, but no procedures should be directly defined in any of the existing classes. *Hint*: Take inspiration from the Self programming language in using *parent* fields.

12.3 *A different task.* Write one of the tasks proposed in the Prologue using this style.

## CHAPTER 13

Closed Maps

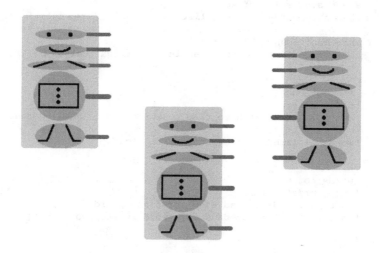

13.1 CONSTRAINTS

▷ The larger problem is decomposed into *things* that make sense for the problem domain.

▷ Each *thing* is a map from keys to values. Some values are procedures/ functions.

▷ The procedures/functions close on the map itself by referring to its slots.

13.2 A PROGRAM IN THIS STYLE

```python
#!/usr/bin/env python
import sys, re, operator, string

# Auxiliary functions that can't be lambdas
#
def extract_words(obj, path_to_file):
    with open(path_to_file) as f:
        obj['data'] = f.read()
    pattern = re.compile('[\W_]+')
    data_str = ''.join(pattern.sub(' ', obj['data']).lower())
    obj['data'] = data_str.split()

def load_stop_words(obj):
    with open('../stop_words.txt') as f:
        obj['stop_words'] = f.read().split(',')
    # add single-letter words
    obj['stop_words'].extend(list(string.ascii_lowercase))

def increment_count(obj, w):
    obj['freqs'][w] = 1 if w not in obj['freqs'] else obj['freqs'][w]+1

data_storage_obj = {
    'data' : [],
    'init' : lambda path_to_file : extract_words(data_storage_obj,
        path_to_file),
    'words' : lambda : data_storage_obj['data']
}

stop_words_obj = {
    'stop_words' : [],
    'init' : lambda : load_stop_words(stop_words_obj),
    'is_stop_word' : lambda word : word in stop_words_obj['stop_words']
}

word_freqs_obj = {
    'freqs' : {},
    'increment_count' : lambda w : increment_count(word_freqs_obj,
        w),
    'sorted' : lambda : sorted(word_freqs_obj['freqs'].items(),
        key=operator.itemgetter(1), reverse=True)
}

data_storage_obj['init'](sys.argv[1])
stop_words_obj['init']()

for w in data_storage_obj['words']():
    if not stop_words_obj['is_stop_word'](w):
        word_freqs_obj['increment_count'](w)

word_freqs = word_freqs_obj['sorted']()
for (w, c) in word_freqs[0:25]:
    print(w, '-', c)
```

13.3 COMMENTARY

THIS STYLE takes yet a different perspective on the Things style explained in the previous chapters. The application is divided in exactly the same way. However, these *things* (*aka* objects) are simple maps from keys to values. Some of these values are simple data, while others are procedures or functions.

Let's take a look at the example program. The program uses essentially the same entities as in the previous 2 examples, but implements those entities in a very different way. Starting in line #22, we have our *objects*:

- data_storage_obj (lines #22–26) models the data storage, similar to what we have seen before. Here we have a dictionary (a hash map) from keywords to values. The first entry, data (line #23), is used to hold the words from the input file. The second one, init (line #24), is our constructor function, or the function that is meant to be called before any others on this dictionary – it simply calls the extract_words procedure, which parses the file and extracts the non-stop words. Note that the init function takes a path to a file as its argument. The third entry on this map is words (line #25), which is mapped to a function returning that object's data field.

- stop_words_obj (lines #28–32) models the stop word manager that we have seen before. Its plain data field, stop_words (line #29), holds the list of stop words. Its init entry is the constructor that fills out the stop_words entry. is_stop_word is a function that returns True if the given argument is a stop word.

- word_freqs_obj (lines #34–38) models the word frequency counter that we have seen before. freqs (line #35) holds the dictionary of word frequencies. increment_count is the procedure that updates the freqs data. The third entry, sorted, is a function returning a sorted list of word-frequency pairs.

In order to see these maps as objects, many conventions need to be followed. First, fields in these objects are the entries that contain simple values, while methods are the entries that contain functions. Constructors are methods that are meant to be called before any other entries. Note also that these simple *objects* refer to themselves in the third person instead of using a self-referential keyword like **this** or **self** – see, for example, line #25. Without additional work to bring in the concept of self-reference, these maps are a lot less expressive than the other concepts of objects that we have seen before.

The rest of the program indexes the right keys at the right time. In lines #40–41, we initialize both data_storage_obj and stop_words_obj. These keys hold procedure values, hence the (...) syntax that denotes a call. Those procedures read the input file and the stop words file, parsing them both into memory. Lines #43–45 loop over the words in data_storage_obj, incrementing the counts in word_freqs_obj for non-stop words. At the end, we request the sorted list (line #47) and print it (lines #48–49).

The *Closed Maps* style illustrates a certain flavor of object-based programming known as *prototypes*. This flavor of OOP is classless: each object is one-of-a-kind. We see this style in JavaScript's concept of object, for example. This style of objects has some interesting possibilities, but also some shortcomings.

On the positive side, it becomes trivial to instantiate objects based on existing ones, for example:

```
>>> data_storage_obj
{'init': <function <lambda> at 0x01E26A70>, 'data': [],
 'words': <function <lambda> at 0x01E26AB0>}
>>> ds2 = data_storage_obj.copy()
>>> ds2
{'init': <function <lambda> at 0x01E26A70>, 'data': [],
 'words': <function <lambda> at 0x01E26AB0>}
```

ds2 is a copy of data_storage_obj at that point in time. From here on, these two objects are relatively independent of each other – although, again, self-referentiality is an issue that would need to be addressed in order to make them truly independent. It is not hard to envision how to create relationship links between related objects using additional slots in the dictionaries.

Extending the objects' functionality at any time is also trivial: we simply need to add more keys to the maps. Removal of keys is also possible.

On the negative side, there is no access control, in the sense that all keys are indexable – there are no hidden keys. It is up to the programmers to use restraint. Also, implementing useful code reuse concepts such as classes, inheritance and delegation requires additional programmer-facing machinery.

But this is a very simple object model that may be useful when the programming languages don't support more advanced notions of objects.

13.4 HISTORICAL NOTES

The idea of objects as prototypes first appeared in the language Self, designed in the late 1980s. Self was heavily inspired by Smalltalk, but deviated from it by using prototypes instead of classes, and delegation instead of inheritance. Self also put forward the idea of objects as collections of "slots." Slots are accessor methods that return values. Self did not use the *Closed Maps* style described here for representing objects, and access to slots of all kinds – simple values as well as methods – was done via messages, in the *Letterbox* style. But indexing a dictionary via keys can be seen as an act of sending messages to it.

13.5 FURTHER READING

Ungar, D. and Smith, R. (1987). Self: The power of simplicity. *OOPSLA '87*. Also in *Lisp and Symbolic Computation* 4(3).

Synopsis: Self was a very nice object-oriented programming language designed in the sequence of Smalltalk, but with some important differences. Although it never went further than being a research prototype, it added to the community's understanding of objects, and it influenced languages such as JavaScript and Ruby.

13.6 GLOSSARY

Prototype: An object in a classless object-oriented language. Prototypes carry with them their own data and functions, and can be changed at any time without affecting others. New prototypes are created by cloning existing ones.

13.7 EXERCISES

13.1 *Another language.* Implement the example program in another language, but preserve the style.

13.2 *Add method.* Delete the last three lines of the example program, and replace the printing of the information with the following. Add a new method to the `word_freqs_obj` called `top25` which sorts its `freqs` data and prints the top 25 entries. Then call that method. Constraint: The program cannot be changed at all until line #46 – your additions should come after that line.

13.3 **this**. In the example program, the prototype objects don't use **this** or **self** to refer to themselves, and, instead, refer to themselves in the third person – e.g. line #25, `data_storage_obj`. Propose another representation of *Closed Maps* that uses the self-reference **this**. So, for example, the method `words` of the data storage object would be
`'words' : lambda : this['data']`

13.4 *Constructor.* In the example program, there is nothing special about constructors. Use your richer representation of object from the previous question to execute the constructor methods every time an object is created.

13.5 *Object combinations.* Let's bring back the methods `info` of Chapter 11. Show how to reuse and overwrite method collections in the *Closed Maps* style by defining a map called `tf_exercise` that contains a generic `info` method that is then reused and overwritten by all the objects of the example program (or your version above).

13.6 *A different task.* Write one of the tasks proposed in the Prologue using this style.

## CHAPTER 14

Abstract Things

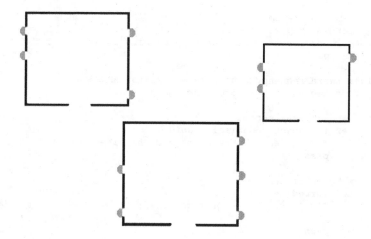

14.1 CONSTRAINTS

▷ The larger problem is decomposed into *abstract things* that make sense for the problem domain.

▷ Each *abstract thing* is described by what operations the things of that abstraction can eventually do.

▷ Concrete things are then bound, somehow, to the abstractions; mechanisms for doing that vary.

▷ The rest of the application uses the things, not by what they are, but by what they do in the abstract.

14.2   A PROGRAM IN THIS STYLE

```
 1  #!/usr/bin/env python
 2  import abc, sys, re, operator, string
 3
 4  #
 5  # The abstract things
 6  #
 7  class IDataStorage (metaclass=abc.ABCMeta):
 8      """ Models the contents of the file """
 9
10      @abc.abstractmethod
11      def words(self):
12          """ Returns the words in storage """
13          pass
14
15  class IStopWordFilter (metaclass=abc.ABCMeta):
16      """ Models the stop word filter """
17
18      @abc.abstractmethod
19      def is_stop_word(self, word):
20          """ Checks whether the given word is a stop word """
21          pass
22
23  class IWordFrequencyCounter(metaclass=abc.ABCMeta):
24      """ Keeps the word frequency data """
25
26      @abc.abstractmethod
27      def increment_count(self, word):
28          """ Increments the count for the given word """
29          pass
30
31      @abc.abstractmethod
32      def sorted(self):
33          """ Returns the words and their frequencies, sorted by
                  frequency"""
34          pass
35
36  #
37  # The concrete things
38  #
39  class DataStorageManager:
40      _data = ''
41      def __init__(self, path_to_file):
42          with open(path_to_file) as f:
43              self._data = f.read()
44          pattern = re.compile('[\W_]+')
45          self._data = pattern.sub(' ', self._data).lower()
46          self._data = ''.join(self._data).split()
47
48      def words(self):
49          return self._data
50
51  class StopWordManager:
52      _stop_words = []
53      def __init__(self):
```

```
54          with open('../stop_words.txt') as f:
55              self._stop_words = f.read().split(',')
56          self._stop_words.extend(list(string.ascii_lowercase))
57
58      def is_stop_word(self, word):
59          return word in self._stop_words
60
61  class WordFrequencyManager:
62      _word_freqs = {}
63
64      def increment_count(self, word):
65          if word in self._word_freqs:
66              self._word_freqs[word] += 1
67          else:
68              self._word_freqs[word] = 1
69
70      def sorted(self):
71          return sorted(self._word_freqs.items(), key=operator.
               itemgetter(1), reverse=True)
72
73
74  #
75  # The wiring between abstract things and concrete things
76  #
77  IDataStorage.register(subclass=DataStorageManager)
78  IStopWordFilter.register(subclass=StopWordManager)
79  IWordFrequencyCounter.register(subclass=WordFrequencyManager)
80
81  #
82  # The application object
83  #
84  class WordFrequencyController:
85      def __init__(self, path_to_file):
86          self._storage = DataStorageManager(path_to_file)
87          self._stop_word_manager = StopWordManager()
88          self._word_freq_counter = WordFrequencyManager()
89
90      def run(self):
91          for w in self._storage.words():
92              if not self._stop_word_manager.is_stop_word(w):
93                  self._word_freq_counter.increment_count(w)
94
95          word_freqs = self._word_freq_counter.sorted()
96          for (w, c) in word_freqs[0:25]:
97              print(w, '-', c)
98
99  #
100 # The main function
101 #
102 WordFrequencyController(sys.argv[1]).run()
```

14.3 COMMENTARY

IN THIS STYLE, the problem is first divided as collections of operations on some abstract data that are important for the problem. These abstract operations are defined in terms of their names and what arguments they receive and return; collectively, they define the access to the data structure that they model. In this first stage, no *concrete things* exist, only the *abstract things*. Any part of the application that uses the data needs only to know about their abstract definition via the operations. In a second stage, concrete implementations are given, which are bound to the *abstract things*. From the point of view of callers, concrete implementations can be replaced with other concrete implementations as long as they provide the same abstract operations.

The Abstract Things style shares some similarities with the Things style, and in several mainstream programming languages, they co-exist.

The example program uses the same entities as in the Things style example: a DataStorage entity, a StopWord entity, a WordFrequency entity, and the WordFrequencyController that starts it all. But the three main data structures are modeled in terms of Abstract Things in lines #7–34. We use Python's Abstract Base Class (ABC) facility as the mechanism to define these abstract things. Three ABCs are defined: IDataStorage (lines #7–13), IStopWordFilter (lines #15–21) and IWordFrequencyCounter (lines #23–34). IDataStorage provides an abstract words operation (lines #11–13); IStopWordFilter provides an abstract is_stop_word operation (lines #19–21); IWordFrequencyCounter provides two abstract operations: increment_count (lines #27–29) and sorted (lines #32–34). Any implementations of these *abstract things* will have to provide concrete implementations of these operations.

Concrete implementations follow in lines #39–71. We use classes as the mechanism to implement concrete data structures accessible via procedures. These classes are identical to the ones in the Things style example, so no explanation is necessary. The important thing to note is that, in this particular implementation, nothing in the classes associates them with the ABCs defined above. The association is done dynamically via the register method of the ABCs (lines #77–79).[1]

The Abstract Things style is often used in conjunction with strong types. For example, both Java and C# support the Abstract Things style through the concept of *interfaces*. In a strongly typed language, the Abstract Things concept encourages program design where the *is-a* relationship is detached from concrete code reuse. Interfaces are used for enforcing the types of expected arguments and return values without having to use concrete implementations (classes).

In contrast to statically typed languages, nothing in the example verifies that entities are of certain abstract (or concrete) types, because Python is

---

[1] register is a method available to all abstract base classes in Python that dynamically associates abstract base classes with any other classes.

dynamically typed. However, the following decorator could be used to add runtime type checking in certain method and constructor calls:

```
1  #
2  # Decorator for enforcing types of arguments in method calls
3  #
4  class AcceptTypes():
5      def __init__(self, *args):
6          self._args = args
7
8      def __call__(self, f):
9          def wrapped_f(*args):
10             for i in range(len(self._args)):
11                 if self._args[i] == 'primitive' and type(args[i
                       +1]) in (str, int, float, bool):
12                     continue
13                 if not isinstance(args[i+1], globals()[self._args[
                       i]]):
14                     raise TypeError("Wrong type")
15
16             f(*args)
17         return wrapped_f
18
19 #
20 # Example of use
21 #
22 class DataStorageManager:
23     # Annotation for type checking
24     @AcceptTypes('primitive', 'IStopWordFilter')
25     def __init__(self, path_to_file, word_filter):
26         with open(path_to_file) as f:
27             self._data = f.read()
28         self._stop_word_filter = word_filter
29         self.__filter_chars_normalize()
30         self.__scan()
31
32     def words(self):
33         return [w for w in self._data if not self.
               _stop_word_filter.is_stop_word(w)]
34
35 ...
36
37 #
38 # The main function creates the objects
39 #
40 stop_word_manager = StopWordManager()
41 storage = DataStorageManager(sys.argv[1], stop_word_manager)
42 word_freq_counter = WordFrequencyManager()
43 WordFrequencyController(storage, word_freq_counter).run()
```

The class AcceptTypes (lines #4–17) is meant to be used as *decorator*. In Python, a decorator is a class whose constructor (__init__) and method (__call__) are called automatically upon the declaration and invocation of the functions that they decorate. Decorations are done using a special @ symbol.

Let's look at line #24, where an `AcceptTypes` decorator is placed right before the definition of the constructor for class `DataStorageManager`. Because of the decorator declaration in line #24, when the Python interpreter first encounters the constructor definition in line #25, it creates an instance of the `AcceptTypes` class, invoking that instance's __init__ constructor. In this case, this constructor (lines #5–6) simply stores the parameter declarations that we have given it – `primitive` for the first argument and `IStopWordFilter` for the second. Later on, when an instance of `DataStorageManager` is created in line #41, and just before the __init__ constructor for that class is actually called, the __call__ method of the decorator `AcceptTypes` is also called (lines #8–17). In this case, our method checks that the arguments provided for the constructor of `DataStorageManager` are of the types that we have declared them to be.

14.4 THIS STYLE IN SYSTEMS DESIGN

The Abstract Things concept plays an important role in large-scale systems design. Software components that use other components, possibly developed by third parties, are often designed against an abstract definition of those components rather than against any concrete implementation. The realization of those abstract interfaces varies depending on the programming language involved.

The Adapter design pattern is an example of a system-level practice that has the same intention as Abstract Things. For example, an application that uses Bluetooth devices will likely use its own adapter as the primary interface to the Bluetooth functionality in order to shield itself from the variability in Bluetooth APIs; a 3D game that supports physics will likely want to use its own adapter to be able to use different physics engines. These adapters typically consist of an interface, or an abstract class, that is then implemented by different concrete classes, each one tailored to interface with each specific third-party library. Such an adapter is playing the same role that an abstract thing plays in small-scale program design: it is shielding the rest of the application from concrete implementations of required functionality.

14.5 HISTORICAL NOTES

The Abstract Things style of programming started emerging in the early 1970s, around the same time that OOP languages were being designed. The original design by Barbara Liskov already included parameterized types, i.e. abstract data types that, internally, use values whose types are variables (e.g. list< T >, where T can be any type).

Many modern programming languages include the concept of abstract things in some form. Java and C# have them in the form of interfaces, which can be parameterized on types. Haskell, a strongly typed pure functional language, has them in the form of type classes. C++ has abstract classes; along with templates, C++'s Standard Template Library (STL) effectively simulates parameterized abstract things.

14.6 FURTHER READING

Cook, W. (2009). On understanding data abstraction, revisited. *Proceedings of the Twenty-Fourth ACM SIGPLAN Conference on Object Oriented Programming Systems Languages and Applications* (OOPSLA '09). ACM, New York, pp. 557–572.
Synopsis: With so many concepts related to objects, it's easy to get confused. William Cook analyzes the subtle but important differences between objects and abstract data types.

Liskov, B. and Zilles, S. (1974) Programming with abstract data types. *Proceedings of the ACM SIGPLAN Symposium on Very High Level Languages,* ACM, New York, pp. 50–59.
Synopsis: The original description of Abstract Data Types, the grandparents of Java and C# interfaces.

14.7 GLOSSARY

Abstract data type: An entity defined in abstract by the operations that it provides.

Decorator: In Python, a *decorator* is a linguistic counterpart to the object-oriented design pattern with the same name, designed to allow behavior to be added to individual objects. A Python decorator allows us to alter functions and methods without changing their source code.

14.8 EXERCISES

14.1 *Another language.* Implement the example program in another language, but preserve the style.

14.2 *Mismatch.* What happens if the sorted method of WordFrequency-Manager is renamed to sorted_freqs, for example? Explain the result in detail.

14.3 *Type checks.* Use the decorator presented in this chapter in order to check the types of parameters passed to certain constructors and methods. Feel free to refactor the original example program in order to make type checking more meaningful. Turn in two versions of your new program, one that type checks and one that fails type checking.

14.4 *A different bind.* Quoting from the description of the example: "In this particular implementation, nothing in the classes associates them with the ABCs defined above. The association is done dynamically via the register method of the ABCs." Do the association between ABCs and concrete implementations in a different way.

14.5 *A different task.* Write one of the tasks proposed in the Prologue using this style.

## CHAPTER 15

Hollywood

15.1 CONSTRAINTS

▷ Larger problem is decomposed into entities using some form of abstraction (objects, modules, or similar).

▷ The entities are never called on directly for actions.

▷ The entities provide interfaces for other entities to be able to register callbacks.

▷ At certain points of the computation, the entities call on the other entities that have registered for callbacks.

15.2   A PROGRAM IN THIS STYLE

```
1  #!/usr/bin/env python
2  import sys, re, operator, string
3
4  #
5  # The "I'll call you back" Word Frequency Framework
6  #
7  class WordFrequencyFramework:
8      _load_event_handlers = []
9      _dowork_event_handlers = []
10     _end_event_handlers = []
11
12     def register_for_load_event(self, handler):
13         self._load_event_handlers.append(handler)
14
15     def register_for_dowork_event(self, handler):
16         self._dowork_event_handlers.append(handler)
17
18     def register_for_end_event(self, handler):
19         self._end_event_handlers.append(handler)
20
21     def run(self, path_to_file):
22         for h in self._load_event_handlers:
23             h(path_to_file)
24         for h in self._dowork_event_handlers:
25             h()
26         for h in self._end_event_handlers:
27             h()
28
29  #
30  # The entities of the application
31  #
32  class DataStorage:
33      """ Models the contents of the file """
34      _data = ''
35      _stop_word_filter = None
36      _word_event_handlers = []
37
38      def __init__(self, wfapp, stop_word_filter):
39          self._stop_word_filter = stop_word_filter
40          wfapp.register_for_load_event(self.__load)
41          wfapp.register_for_dowork_event(self.__produce_words)
42
43      def __load(self, path_to_file):
44          with open(path_to_file) as f:
45              self._data = f.read()
46          pattern = re.compile('[\W_]+')
47          self._data = pattern.sub(' ', self._data).lower()
48
49      def __produce_words(self):
50          """ Iterates through the list words in storage
51              calling back handlers for words """
52          data_str = ''.join(self._data)
53          for w in data_str.split():
54              if not self._stop_word_filter.is_stop_word(w):
```

```
55              for h in self._word_event_handlers:
56                  h(w)
57
58      def register_for_word_event(self, handler):
59          self._word_event_handlers.append(handler)
60
61  class StopWordFilter:
62      """ Models the stop word filter """
63      _stop_words = []
64      def __init__(self, wfapp):
65          wfapp.register_for_load_event(self.__load)
66
67      def __load(self, ignore):
68          with open('../stop_words.txt') as f:
69              self._stop_words = f.read().split(',')
70          # add single-letter words
71          self._stop_words.extend(list(string.ascii_lowercase))
72
73      def is_stop_word(self, word):
74          return word in self._stop_words
75
76  class WordFrequencyCounter:
77      """ Keeps the word frequency data """
78      _word_freqs = {}
79      def __init__(self, wfapp, data_storage):
80          data_storage.register_for_word_event(self.
81              __increment_count)
81          wfapp.register_for_end_event(self.__print_freqs)
82
83      def __increment_count(self, word):
84          if word in self._word_freqs:
85              self._word_freqs[word] += 1
86          else:
87              self._word_freqs[word] = 1
88
89      def __print_freqs(self):
90          word_freqs = sorted(self._word_freqs.items(), key=operator
                  .itemgetter(1), reverse=True)
91          for (w, c) in word_freqs[0:25]:
92              print(w, '-', c)
93
94  #
95  # The main function
96  #
97  wfapp = WordFrequencyFramework()
98  stop_word_filter = StopWordFilter(wfapp)
99  data_storage = DataStorage(wfapp, stop_word_filter)
100 word_freq_counter = WordFrequencyCounter(wfapp, data_storage)
101 wfapp.run(sys.argv[1])
```

15.3 COMMENTARY

THIS STYLE differs from the previous ones by its use of inversion of control: rather than an entity e_1 calling another entity e_2 with the purpose of getting some information, e_1 registers with e_2 for a callback; e_2 then calls back e_1 at a later time.

The example program is divided into entities that are very similar to the previous styles: there is an entity for data storage (DataStorage – lines #32–59), another one for dealing with stop words (StopWordFilter – lines #61–74), and a third one for managing the word frequency pairs (WordFrequencyCounter – lines #76–92). Additionally, we define a WordFrequencyFramework entity (lines #7–27), which is responsible for orchestrating the execution of the program.

Let's start by analyzing the WordFrequencyFramework. This class provides three registration methods and a fourth method, run, that executes the program. The run method (lines #21–27) tells the story of this class: the application has been decomposed into three phases, namely a load phase, a dowork phase and an end phase; other entities of the application register for callbacks for each of those phases by calling register_for_load_event (lines #12–13), register_for_dowork_event (lines #15–16), and register_for_end_event (lines #18–19), respectively. The corresponding handlers are then called by the run procedure at the right times. Figuratively speaking, WordFrequencyFramework is like a puppet master pulling the strings on the application objects below so that they actually do what they have to do at specific times.

Next, let's look at the three application classes, and how they use the WordFrequencyFramework, and each other.

As in previous examples, DataStorage models the input data. The way this example was designed, this class produces *words events* that other entities can register for. As such, it provides an event registration method, register_for_word_event (lines #58–59). Besides that, this class's constructor (lines #38–41) gets a reference to the StopWordFilter object (more on this later), and then registers with the WordFrequencyFramework for two events: load and dowork. On load events, this class opens and reads the entire contents of the input file, filtering the characters and normalizing them to lowercase (lines #43–47); on dowork events, this class splits the data into words (line #53), and then, for every non-stop word, it calls the handlers of entities that have registered for *word* events (lines #53–56).

The constructor of StopWordFilter (lines #64–65) registers with WordFrequencyFramework for the load event. When that handler is called back, it simply opens the stop words file and produces the list of all stop words (lines #67–71). This class exposes the method is_stop_word (lines #73–74) which can be called by other classes – in this case, this method is called only by `DataStorage` (line #54) when it iterates through its list of words from the input file.

The `WordFrequencyCounter` keeps a record of the word-count pairs. Its constructor (lines #79–81) gets the reference for the `DataStorage` entity and registers with it a handler for *word* events (line #80) – remember, `DataStorage` calls back these handlers in lines #55–56. It then registers with the `WordFrequencyFramework` for the *end* event. When that handler is called, it prints the information on the screen (lines #89–92).

Let's go back to `WordFrequencyFramework`. As mentioned before, this class acts like a puppet master. Its `run` method simply calls all handlers that have registered for the three phases of the application, *load, dowork* and *end*. In our case, `DataStorage` and `StopWordFilter` both register for the *load* phase (lines #40 and #65, respectively), so their handlers are called when lines #22–23 execute; only `DataStorage` registers for the *dowork* phase (line #41), so only its handler defined in lines #49–56 is called when lines #24–25 execute; finally, only `WordFrequencyCounter` registers for the *end* phase (line #81), so its handler, defined in lines #89–92, is called when lines #26–27 execute.

The Hollywood style of programming seems rather contrived, but it has one interesting property: rather than hardwiring callers to callees at specific points of the program (i.e. function calls, where the binding is done by naming functions), it reverts that relation, allowing a callee to trigger actions in many callers at a time determined by the callee. This supports a different kind of module composition, as many modules can register handlers for the same event on a provider.

This style is used in many object-oriented frameworks, as it is a powerful mechanism for the framework code to trigger actions in arbitrary application code. Inversion of control is precisely what makes frameworks different from regular libraries. The Hollywood style, however, should be used with care, as it may result in code that is extremely hard to understand. We will see variations of this style in subsequent chapters.

15.4   THIS STYLE IN SYSTEMS DESIGN

Inversion of control is an important concept in distributed systems design. It is sometimes useful for a component in one node of the network to ask another component on another node to call back when specific conditions occur, rather than the first component polling the second one periodically. Taken to the extreme, this concept results in event-driven architectures (see Chapter 16).

15.5   HISTORICAL NOTES

The Hollywood style has its origins in asynchronous hardware interrupts. In operating systems design, interrupt handlers play a critical role in ensuring separation between layers. As seen by the example, the Hollywood style doesn't require asynchronicity, although the callbacks can be asynchronous.

This style gained traction in application software during the 1980s in the context of Smalltalk and Graphical User Interfaces.

15.6 FURTHER READING

Johnson, R. and Foote, B. (1988). Designing reusable classes. *Journal of Object-Oriented Programming* 1(2): 22–35.
Synopsis: The first written account of the idea of inversion of control, which emerged in the context of Smalltalk.

Fowler, M. (2005). InversionOfControl. Blog post at:
http://martinfowler.com/bliki/InversionOfControl.html
Synopsis: Martin Fowler gives a short and sweet description of inversion of control.

15.7 GLOSSARY

Inversion of control: Any technique that supports independently developed code being called by a generic library or component.

Framework: A special kind of library, or reusable component, providing a generic application functionality that can be customized with additional user-written code.

Handler: A function that is to be called back at a later time.

15.8 EXERCISES

15.1 *Another language.* Implement the example program in another language, but preserve the style.

15.2 *Words with z.* Change the given example program so that it implements an additional task: after printing out the list of 25 top words, it should print out the number of non-stop words with the letter z. Additional constraints: (i) no changes should be made to the existing classes; adding new classes and more lines of code to the main function is allowed; (ii) files should be read only once for both term frequency and "words with z" tasks.

15.3 *Words with z in other styles.* Consider all the previous styles. For each of them, try to do the additional "words with z" task observing the constraints stated above. If you are able to do it, show the code; if not, explain why you think it can't be done.

15.4 *A different task.* Write one of the tasks proposed in the Prologue using this style.

## CHAPTER 16

Bulletin Board

16.1 CONSTRAINTS

▷ Larger problem is decomposed into entities using some form of abstraction (objects, modules, or similar).

▷ The entities are never called on directly for actions.

▷ Existence of an infrastructure for publishing and subscribing to events, *aka* the bulletin board.

▷ Entities post event subscriptions (*aka* "wanted") to the bulletin board and publish events (*aka* "offered") to the bulletin board. The bulletin board infrastructure does all the event management and distribution.

16.2 A PROGRAM IN THIS STYLE

```python
1  #!/usr/bin/env python
2  import sys, re, operator, string
3
4  #
5  # The event management substrate
6  #
7  class EventManager:
8      def __init__(self):
9          self._subscriptions = {}
10
11     def subscribe(self, event_type, handler):
12         if event_type in self._subscriptions:
13             self._subscriptions[event_type].append(handler)
14         else:
15             self._subscriptions[event_type] = [handler]
16
17     def publish(self, event):
18         event_type = event[0]
19         if event_type in self._subscriptions:
20             for h in self._subscriptions[event_type]:
21                 h(event)
22
23 #
24 # The application entities
25 #
26 class DataStorage:
27     """ Models the contents of the file """
28     def __init__(self, event_manager):
29         self._event_manager = event_manager
30         self._event_manager.subscribe('load', self.load)
31         self._event_manager.subscribe('start', self.produce_words)
32
33     def load(self, event):
34         path_to_file = event[1]
35         with open(path_to_file) as f:
36             self._data = f.read()
37         pattern = re.compile('[\W_]+')
38         self._data = pattern.sub(' ', self._data).lower()
39
40     def produce_words(self, event):
41         data_str = ''.join(self._data)
42         for w in data_str.split():
43             self._event_manager.publish(('word', w))
44         self._event_manager.publish(('eof', None))
45
46 class StopWordFilter:
47     """ Models the stop word filter """
48     def __init__(self, event_manager):
49         self._stop_words = []
50         self._event_manager = event_manager
51         self._event_manager.subscribe('load', self.load)
52         self._event_manager.subscribe('word', self.is_stop_word)
53
54     def load(self, event):
```

```
55          with open('../stop_words.txt') as f:
56              self._stop_words = f.read().split(',')
57          self._stop_words.extend(list(string.ascii_lowercase))
58
59      def is_stop_word(self, event):
60          word = event[1]
61          if word not in self._stop_words:
62              self._event_manager.publish(('valid_word', word))
63
64  class WordFrequencyCounter:
65      """ Keeps the word frequency data """
66      def __init__(self, event_manager):
67          self._word_freqs = {}
68          self._event_manager = event_manager
69          self._event_manager.subscribe('valid_word', self.
                increment_count)
70          self._event_manager.subscribe('print', self.print_freqs)
71
72      def increment_count(self, event):
73          word = event[1]
74          if word in self._word_freqs:
75              self._word_freqs[word] += 1
76          else:
77              self._word_freqs[word] = 1
78
79      def print_freqs(self, event):
80          word_freqs = sorted(self._word_freqs.items(), key=operator
                .itemgetter(1), reverse=True)
81          for (w, c) in word_freqs[0:25]:
82              print(w, '-', c)
83
84  class WordFrequencyApplication:
85      def __init__(self, event_manager):
86          self._event_manager = event_manager
87          self._event_manager.subscribe('run', self.run)
88          self._event_manager.subscribe('eof', self.stop)
89
90      def run(self, event):
91          path_to_file = event[1]
92          self._event_manager.publish(('load', path_to_file))
93          self._event_manager.publish(('start', None))
94
95      def stop(self, event):
96          self._event_manager.publish(('print', None))
97
98  #
99  # The main function
100 #
101 em = EventManager()
102 DataStorage(em), StopWordFilter(em), WordFrequencyCounter(em)
103 WordFrequencyApplication(em)
104 em.publish(('run', sys.argv[1]))
```

16.3 COMMENTARY

THIS STYLE is a logical end point of the previous style, where components never call each other directly. Furthermore, the infrastructure that binds the entities together is made even more generic by removing any application-specific semantics and adopting only two generic operations: publish an event and subscribe to an event type.

The example program defines an EventManager class that implements the generic bulletin board concept (lines #7–21). This class wraps a dictionary of subscriptions (line #9), and has two methods:

- subscribe (lines #11–15) takes an event type and a handler, and it appends the handler to the subscription dictionary using the event type as key.

- publish (lines #17–21) takes an event, which may be a complex data structure. In our case, this data structure is assumed to have the event type in its first position (line #18). It then proceeds to call all handlers that have been registered for that event type (lines #19–21).

The entities of the example program are similar to those of the previous styles: data storage (lines #26–44), stop word filter (lines #46–62) and word frequency counter (lines #64–82). Additionally, the WordFrequencyApplication class (lines #84–96) starts and ends the word frequency application. These classes use EventManager in order to interact with each other, by requesting notifications of events and publishing their own events. The classes are arranged more or less in a pipeline of events, as follows:

The application starts with the main function generating the run event (line #104), which is handled by WordFrequencyApplication (line #87). As part of the reaction to the run event, WordFrequencyApplication first triggers the load event (line #92) that has actions in both DataStorage (line #30) and StopWordFilter (line #51), resulting in files being read and processed; then, it triggers the start event (line #93) that has action in DataStorage (line #31), resulting in an iteration over words; from then on, for each word, DataStorage triggers word events (line #43) that have actions in StopWordFilter (line #52); in turn, in the presence of a non-stop word, StopWordFilter triggers valid_word events (line #62) that have actions in WordFrequencyCounter (line #69), resulting in incremented counters. When there are no more words, DataStorage triggers the eof event (line #44) which has actions in WordFrequencyApplication (line #88), resulting in the final information being printed on the screen.

In the example program, events are implemented as tuples having the event type as string in the first position and any additional arguments as subsequent elements of the tuple. So, for example, the run event generated by the main function (line #104) is ('run', sys.argv[1]), the word event generated by DataStorage (line #43) is ('word', w), etc.

The Bulletin Board style is often used with asynchronous components, but as seen here, that is not required. The infrastructure for handling events may be as simple as the one shown here or much more sophisticated, with several components interacting for the distribution of events. The infrastructure may also include more sophisticated event structures that support more detailed event filtering – for example, rather than simple subscriptions to event types, as shown here, components may subscribe to a combination of event types and contents.

Like the previous style, the Bulletin Board style supports inversion of control, but taken to its most extreme and minimal form – events generated by some components in the system may cause actions in other components of the system. The subscription is anonymous, so a component generating an event, in principle, doesn't know all the components that are going to handle that event. This style supports a very flexible entity composition mechanism (via events), but, like the previous style, in certain cases it may lead to systems whose erroneous behaviors are difficult to trace.

16.4 THIS STYLE IN SYSTEMS DESIGN

This style is at its best as an architecture for distributed systems known as publish-subscribe. Publish-subscribe architectures are popular in companies with large computational infrastructures, because they are very extensible and support unforeseen system evolution – components can be easily added and removed, new types of events can be distributed, etc.

16.5 HISTORICAL NOTES

Historically, this style can be traced to the USENET, a distributed news system developed in the late 1970s. The USENET was, indeed, the first electronic bulletin board, where users could post (*publish*) and read articles via *subscription* to specific news channels. Unlike many modern pub-sub systems, the USENET was truly distributed, in the sense that there was no central server for managing news; instead, the system consisted of loosely connected news servers, which could be hosted by different organizations, and which distributed the users' posts among themselves.

The USENET was one particular kind of distributed system – one for sharing user-generated news. With the advent of the Web, USENET became less and less popular, but the concept had a second life in RSS, a protocol that enables publishers of Web content to notify subscribers of that content.

Over the years, the bulletin board concept has seen applications in many other areas. In the 1990s, there was a considerable amount of work in generalizing the concept to all sorts of distributed system infrastructures.

16.6 FURTHER READING

Oki, B., Pfluegl, M., Siegel, A. and Skeen, D. (1993). The Information Bus: An architecture for extensible systems. *ACM SIGOPS* 27(5): 58–68. *Synopsis*: One of the earliest written accounts of the idea of publish-subscribe.

Truscott, T. (1979). Invitation to a General Access UNIX* Network. Fax of first official announcement of the USENET. Available at http://www.newsdemon.com/first-official-announcement-usenet.php *Synopsis*: Long before Facebook and Hacker News, there was the Usenet and its many newsgroups to which people subscribed and posted. The Usenet was the ultimate electronic bulletin board.

16.7 GLOSSARY

Event: A data structure produced by a component at a certain point in time and meant to be distributed to other components waiting for it.

Publish: An operation provided by the event distribution infrastructure allowing components to distribute events to other components.

Subscribe: An operation provided by the event distribution infrastructure allowing components to express their interest in specific kinds of events.

16.8 EXERCISES

16.1 *Another language.* Implement the example program in another language, but preserve the style.

16.2 *Words with z.* Change the given example program so that it implements an additional task: after printing out the list of 25 top words, it should print out the number of non-stop words with the letter z. Additional constraints: (i) no changes should be made to the existing classes; adding new classes and more lines of code to the main function is allowed; (ii) files should be read only once for both term frequency and "words with z" tasks.

16.3 *Unsubscribe.* Pub-sub architectures usually also support the concept of unsubscribing from event types. Change the example program so that EventManager supports the operation unsubscribe. Make the components unsubscribe from event types at appropriate times. Show that your unsubscription mechanism works correctly.

16.4 *A different task.* Write one of the tasks proposed in the Prologue using this style.

V

Reflection and Metaprogramming

We have seen styles that use functions, procedures and objects; we have also seen functions and objects being passed around and stored in variables, as regular data values. However, these programs are blind tools: we see them, and they interact with us via input/output, but they don't see themselves. This part of the book contains a few styles related to the use of computational reflection and metaprogramming. Reflection is about programs being somehow aware of themselves. Metaprogramming is programming that involves the program accessing, and even changing, itself as it executes. Besides its geeky appeal, metaprogramming can be very useful for engineering software systems that evolve over time.

Reflection falls into the category of programming concepts that are too powerful for their own sake, and that, therefore, should be used with great care. However, many modern composition techniques would not be possible without it.

## CHAPTER 17

Introspective

17.1 CONSTRAINTS

▷ The problem is decomposed using some form of abstraction (procedures, functions, objects, etc.).

▷ The abstractions have access to information about themselves and others, although they cannot modify that information.

17.2   A PROGRAM IN THIS STYLE

```python
#!/usr/bin/env python
import sys, re, operator, string, inspect

def read_stop_words():
    """ This function can only be called from a function
        named extract_words."""
    # Meta-level data: inspect.stack()
    if inspect.stack()[1][3] != 'extract_words':
        return None

    with open('../stop_words.txt') as f:
        stop_words = f.read().split(',')
    stop_words.extend(list(string.ascii_lowercase))
    return stop_words

def extract_words(path_to_file):
    # Meta-level data: locals()
    with open(locals()['path_to_file']) as f:
        str_data = f.read()
    pattern = re.compile('[\W_]+')
    word_list = pattern.sub(' ', str_data).lower().split()
    stop_words = read_stop_words()
    return [w for w in word_list if not w in stop_words]

def frequencies(word_list):
    # Meta-level data: locals()
    word_freqs = {}
    for w in locals()['word_list']:
        if w in word_freqs:
            word_freqs[w] += 1
        else:
            word_freqs[w] = 1
    return word_freqs

def sort(word_freq):
    # Meta-level data: locals()
    return sorted(locals()['word_freq'].items(), key=operator.
        itemgetter(1), reverse=True)

def main():
    word_freqs = sort(frequencies(extract_words(sys.argv[1])))
    for (w, c) in word_freqs[0:25]:
        print(w, '-', c)

if __name__ == "__main__":
    main()
```

17.3 COMMENTARY

THE FIRST STAGE toward computational reflection requires that the programs have access to information about themselves. The ability for a program to access information about itself is called *introspection*. Not all programming languages support introspection, but some do. Python, Java, C#, Ruby, JavaScript, and PHP are examples of languages that support it; C and C++ are examples of languages that don't support it.

The example program uses just a small amount of introspection, enough to illustrate the main concept. The first encounter with introspection is in line #8: the read_stop_words function checks who its caller function is, and returns no value for all callers except the function extract_words. This is a somewhat draconian pre-condition for this function, but checking who callers are may make sense in certain situations, and it can only be done in a language that exposes the call stack to programs. Access to the identification of the caller is done by inspecting the call stack (inspect.stack()), accessing the previous frame ([1], with [0] being the current frame) and accessing its third element, which is the function name.

The other occurrences of introspection are all the same: accessing arguments passed to functions via an introspective runtime structure, locals() – see lines #18, #28, #37. Normally, arguments are referenced directly by name; for example, in line #18 one would normally write:

```
def extract_words(path_to_file):
  with open(path_to_file) as f:
    ...
```

Instead, we are accessing it via locals()['path_to_file']. In Python, locals() is a function that returns a dictionary representing the current local symbol table. One can iterate through this symbol table to find out all the local variables available to a function. It's not uncommon to find its use in idioms such as this one:

```
def f(a, b):
  print "a is %(a)s, b is %(b)s" % locals()
```

where *a* and *b* within the string serve as indexes to the local variables dictionary.

Python has powerful introspective functions, some of them built-in (e.g. callable, which checks whether a given value is a callable entity such as a function) and others provided by modules such as the inspect module. Other languages that support introspection provide similar facilities, even if the APIs are very different. These facilities open the door to an entire new dimension of program design, one that takes the program itself into account.

17.4 THIS STYLE IN SYSTEMS DESIGN

When used with care in justified situations, accessing the program's internal structures for getting additional context can enable powerful behaviors with relatively low programming complexity. However, the use of introspection adds an additional indirection to programs that is not always desirable, and that may make the programs hard to understand. Introspection should be avoided, unless the alternatives are worse.

17.5 GLOSSARY

Introspection: The ability for a program to access information about itself.

17.6 EXERCISES

17.1 *Another language.* Implement the example program in another language, but preserve the style.

17.2 *Print out information.* Change the example program so that it prints out the following information in the beginning of each function:

```
My name is <function name>
    my locals are <k1=v1, k2=v2, k3=v3, ...>
    and I'm being called from  <name of
    caller function>
```

Additional constraint: these messages should be printed as the result of a call to a function named print_info(), with no arguments. So, for example:

```
def read_stop_words():
  print_info()
  ...
```

17.3 *Browse.* Let's back to Chapters 11 and 12. For one of those chapters, using either the Python example code or your own version in another language, add code at the end of the program iterating through the classes of the program using the introspection capabilities of the language. At each iteration your new code should print out the name of the class and the names of the methods.

## CHAPTER 18

Reflective

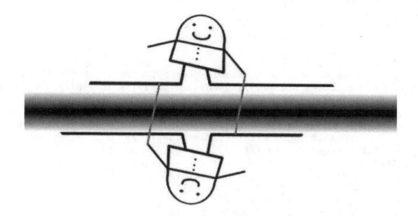

18.1 CONSTRAINTS

▷ The program has access to information about itself, i.e. introspection.

▷ The program can modify itself – adding more abstractions, variables, etc., at runtime.

18.2   A PROGRAM IN THIS STYLE

```python
#!/usr/bin/env python
import sys, re, operator, string, os

#
# Two down-to-earth things
#
stops = set(open("../stop_words.txt").read().split(",") + list(
    string.ascii_lowercase))

def frequencies_imp(word_list):
    word_freqs = {}
    for w in word_list:
        if w in word_freqs:
            word_freqs[w] += 1
        else:
            word_freqs[w] = 1
    return word_freqs

#
# Let's write our functions as strings.
#
if len(sys.argv) > 1:
    extract_words_func = "lambda name : [x.lower() for x in re.
        split('[^a-zA-Z]+', open(name).read()) if len(x) > 0 and x
        .lower() not in stops]"
    frequencies_func = "lambda wl : frequencies_imp(wl)"
    sort_func = "lambda word_freq: sorted(word_freq.items(), key=
        operator.itemgetter(1), reverse=True)"
    filename = sys.argv[1]
else:
    extract_words_func = "lambda x: []"
    frequencies_func = "lambda x: []"
    sort_func = "lambda x: []"
    filename = os.path.basename(__file__)
#
# So far, this program isn't much about term-frequency. It's about
# a bunch of strings that look like functions.
# Let's add our functions to the "base" program, dynamically.
#
exec('extract_words = ' + extract_words_func)
exec('frequencies = ' + frequencies_func)
exec('sort = ' + sort_func)

#
# The main function. This would work just fine:
#   word_freqs = sort(frequencies(extract_words(filename)))
#
word_freqs = locals()['sort'](locals()['frequencies'](locals()['
    extract_words'](filename)))

for (w, c) in word_freqs[0:25]:
    print(w, '-', c)
```

18.3 COMMENTARY

THE SECOND AND FINAL STAGE towards computational reflection requires that the programs be able to modify themselves. The ability for a program to examine and modify itself is called *reflection*. This is an even more powerful proposition than introspection and, as such, of all the languages that support introspection, only a small subset of them support full reflection. Ruby is an example of a language supporting full reflection; Python and JavaScript support it with restrictions; Java and C# support only a small set of reflective operations.

The example program exercises some of Python's reflection facilities. The program starts by reading the stop words file in the normal way (line #8), followed by the definition of a normal function for counting word occurrences that would be too awkward to implement reflectively in Python (lines #7–16).

Next, the main program functions are defined (lines #21–30). But rather than defining them using normal function definitions, we define them at the meta-level: at that level, we have anonymous functions expressed as strings. These are lazy (unevaluated) pieces of program, as lazy as it gets: unprocessed strings whose contents happens to be Python code.

More importantly, the contents of these stringified functions depend on whether the user has provided an input file as argument to the program or not. If there is an input argument, the functions do something useful (lines #21–24); if there isn't, the functions don't do anything, simply returning the empty list (lines #26–29).

Let's look into the three functions defined in lines #22–24:

- In line #22, we have the meta-level definition of a function that extracts words from a file. The file name is given as its only argument, name.

- In line #23, we have the meta-level definition of a function that counts word occurrences given a list of words. In this case, it simply calls the base-level function that we have defined in lines #10–17.

- In line #24, we have the meta-level definition of a function that sorts a dictionary of word frequencies.

At this point of the program, all that exists in the program is: (1) the stops variable that has been defined in line #7; (2) the frequencies_imp function that has been defined in lines #9–16; (3) the three variables extract_words_func, frequencies_func and sort_func, which hold on to strings – those strings are different depending on whether there was an input argument or not.

The next three lines (#36–38) are the part of the program that effectively makes the program change itself. exec is a Python statement that supports dynamic execution of Python code.[1] Whatever is given as argument (a string)

---

[1] Other languages (e.g. Scheme, JavaScript) provide a similar facility through eval.

is assumed to be Python code. In this case we are giving it assignment statements in the form $a = b$, where a is a name (extract_words, frequencies and sort), and b is the variable bound to a stringified function defined in lines #21–30. So, for example, the complete statement in line #37 is either

```
exec('frequencies = lambda wl : frequencies_imp(wl)')
```

or

```
exec('frequencies = lambda x : []')
```

depending on whether there is an input argument given to the program.

exec takes its argument, parses the code, eventually raising exceptions if there are syntax errors, and executes it. After line #38 is executed, the program will contain 3 additional function variables whose values depend on the existence of the input argument.

Finally, line #46 calls those functions. As stated in the comment in lines #41–44, this is a somewhat contrived form of function calling; it is done only to illustrate the lookup of functions via the local symbol table, as explained in the previous chapter.

At this point, the reader should be puzzled about the definition of functions as strings in lines #21–30, followed by their runtime loading via exec in lines #36–38. After all, we could do this instead:

```
1  if len(sys.argv) > 1:
2      extract_words = lambda name : [x.lower() for x in re.split('[^
           a-zA-Z]+', open(name).read()) if len(x) > 0 and x.lower()
           not in stops]
3      frequencies = lambda word_list : frequencies_imp(word_list)
4      sort = lambda word_freq: sorted(word_freq.iteritems(), key=
           operator.itemgetter(1), reverse=True)
5      filename = sys.argv[1]
6  else:
7      extract_words = lambda x: []
8      frequencies = lambda x: []
9      sort = lambda x: []
10     filename = os.path.basename(__file__)
```

Python being a dynamic language with higher-order functions, it supports dynamic definition of functions, as illustrated above. This would achieve the goal of having different function definitions depending on the existence of the input argument, while avoiding reflection (exec and friends) altogether.

18.4   THIS STYLE IN SYSTEMS DESIGN

Indeed, the example program is a bit artificial and begs the question: *when is reflection needed?*

In general, *reflection is needed when the ways by which programs will be modified cannot be predicted at design time.* Consider, for example, the case in which the concrete implementation of the extract_words function in the example would be given by an external file provided by the user. In that case, the designer of the example program would not be able to define the function *a priori*, and the only solution to support such a situation would be to treat the function as string and load it at runtime via reflection. Our example program does not account for that situation, hence the use of reflection here is questionable. In the next two chapters we will see two examples of reflection being used for very good purposes that could not be supported without it.

18.5 HISTORICAL NOTES

Reflection was studied in philosophy and formalized in logic long before being brought into programming. Computational reflection emerged in the 1970s within the LISP world. Its emergence within the LISP community was a natural consequence of early work in artificial intelligence, which, for the first few years, was coupled with work in LISP. At the time, it was assumed that any system that would become intelligent would need to gain awareness of itself – hence the effort in formalizing what such awareness might look like within programming models. Those ideas influenced the design of Smalltalk in the 1980s, which, from early on, supported reflection. Smalltalk went on to influence all OOP languages, so reflection concepts were brought to OOP languages early on. During the 1990s, as the work in artificial intelligence took new directions away from LISP, the LISP community continued the work on reflection; that work's pinnacle was the MetaObject Protocol (MOP) in the Common LISP Object System (CLOS). The software engineering community took notice, and throughout the 1990s there was a considerable amount of work in understanding reflection and its practical benefits. It was clear that the ability to deal with unpredictable changes was quite useful, but dangerous at the same time, and some sort of balance via proper APIs would need to be defined. These ideas found their way to all major programming languages designed since the 1990s.

18.6 FURTHER READING

Demers, F.-N. and Malenfant, J. (1995). Reflection in logic, functional and object-oriented programming: a short comparative study. *IJCAI'95 Workshop on Reflection and Metalevel Architectures and Their Applications in AI.*
Synopsis: A nice retrospective overview of computational reflection in various languages.

Kiczales, G., des Riviere, J. and Bobrow, D. (1991). *The Art of the Metaobject Protocol.* MIT Press. 345 pages.
Synopsis: The Common LISP Object System included powerful reflective and metaprogramming facilities. This book explains how to make objects and their metaobjects work together in CLOS.

Maes, P. (1987). Concepts and Experiments in Computational Reflection. *Object-Oriented Programming Systems, Languages and Applications (OOPSLA'87)*.
Synopsis: Patti Maes brought Brian Smith's ideas to object-oriented languages.

Smith, B. (1984). Reflection and Semantics in LISP. *ACM SIGPLAN Symposium on Principles of Programming Languages (POPL'84)*.
Synopsis: Brian Smith was the first one to formulate computational reflection. He did it in the context of LISP. This is the original paper.

18.7   GLOSSARY

Computational reflection: The ability for programs to access information about themselves and modify themselves.

eval: A function, or statement, provided by several programming languages that evaluates a quoted value (e.g. a string) assumed to be the representation of a program. eval is one of the two foundational pieces of meta-circular interpreters underlying many programming languages, the other one being apply. Any language that exposes eval to programmers is capable of supporting reflection. However, eval is too powerful and often considered harmful. Work on computational reflection focused on how to *tame* eval.

18.8   EXERCISES

18.1   *Another language.* Implement the example program in another language, but preserve the style.

18.2   *From a file.* Modify the example program so that the implementation of extract_words is given by a file. The command line interface should be:

```
$ python tf-16-1.py ../pride-and-prejudice.txt ext1.py
```

Provide at least two alternative implementations of that function (i.e. two files) that make the program work correctly.

18.3   *More reflection.* The example program doesn't use reflection for reading the stop words (line #7) and counting the word occurrences (lines #9–16). Modify the program so that it also uses reflection to do those tasks. If you can't do it, explain what the obstacles are.

18.4   *A different task.* Write one of the tasks proposed in the Prologue using this style.

## CHAPTER 19

Aspects

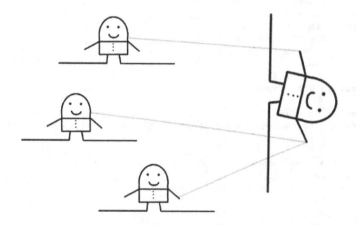

19.1 CONSTRAINTS

▷ The problem is decomposed using some form of abstraction (procedures, functions, objects, etc.).

▷ Aspects of the problem are added to the main program without any edits to the source code of the abstractions or the sites that use them.

▷ An external binding mechanism binds the abstractions with the aspects.

19.2 A PROGRAM IN THIS STYLE

```python
#!/usr/bin/env python
import sys, re, operator, string, time

#
# The functions
#
def extract_words(path_to_file):
    with open(path_to_file) as f:
        str_data = f.read()
    pattern = re.compile('[\W_]+')
    word_list = pattern.sub(' ', str_data).lower().split()
    with open('../stop_words.txt') as f:
        stop_words = f.read().split(',')
    stop_words.extend(list(string.ascii_lowercase))
    return [w for w in word_list if not w in stop_words]

def frequencies(word_list):
    word_freqs = {}
    for w in word_list:
        if w in word_freqs:
            word_freqs[w] += 1
        else:
            word_freqs[w] = 1
    return word_freqs

def sort(word_freq):
    return sorted(word_freq.items(), key=operator.itemgetter(1),
        reverse=True)

# The side functionality
def profile(f):
    def profilewrapper(*arg, **kw):
        start_time = time.time()
        ret_value = f(*arg, **kw)
        elapsed = time.time() - start_time
        print("%s(...) took %s secs" % (f.__name__, elapsed))
        return ret_value
    return profilewrapper

# join points
tracked_functions = [extract_words, frequencies, sort]
# weaver
for func in tracked_functions:
    globals()[func.__name__]=profile(func)

word_freqs = sort(frequencies(extract_words(sys.argv[1])))

for (w, c) in word_freqs[0:25]:
    print(w, '-', c)
```

19.3 COMMENTARY

T HIS STYLE can be described as "restrained reflection" for the specific purpose of injecting arbitrary code before and after designated points of existing programs. One reason for doing that might be not having access to, or not wanting to modify, the source code while wanting to add additional functionality to the program's functions; another reason might be to simplify development by localizing code that is usually scattered throughout the program.

The example program starts by defining the three main program functions: extract_words (lines #7–15), which extract the non-stop words from the input file into a list; frequencies (lines #17–24), which counts the number of occurrences of words on a list; and sort (lines #26–27), which sorts a given word-frequency dictionary. The program could run as is simply by executing lines #45–48.

In addition to the main program, we are adding a side functionality: we want to compute the time that each function takes to execute. This functionality is part of a set of diagnosis actions known as *profiling*. There are many ways of implementing this side functionality. The most straightforward way involves adding a couple of lines of code to each function, in the beginning and in the end. We could also do it outside of the functions, at the calling sites. However, that would violate the constraints of the Aspects style.

One of the constraints of this style is that the aside functionality should bring no edits to the affected functions or their call sites. Given this constraint, the ways of implementing the side functionality narrow down to the use of some form of reflection, i.e. changing the program after the fact. The example program does it as follows.

We define a profile function (lines #30–37) that is a function wrapper: it takes a function argument (f) and returns another function, profilewrapper (line #37), that wraps around the original function f (line #33), adding profiling code before (line #32) and after (lines #34–35); then the wrapper function returns the value that the original function returned (line #36).

The machinery for profiling is in place, but it is still not enough. The last piece that is missing is the expression of our intent about profiling the functions of our program. Again, this can be done in a number of different ways. This style of programming calls for an external binding mechanism: rather than tagging the functions as profileable (e.g. using a decorator), we need to make them profileable without that information being directly attached to them, and, instead, that information being localized in another part of the program.

As such, our program first states which functions should be profiled (line #40); these are called the *join points* between the program's functions and the side functionality. Next, we use full-on reflection: for each of the functions to be profiled, we replace their name's binding in the symbol table with the wrapper function instead. Note that we are changing the program's internal structure: for example, the binding between the name `extract_words` and the corresponding function defined in lines #7–15 has been broken; instead we now have the name `extract_words` bound to an instance of the `profile` function taking the function `extract_words` as a parameter. We changed the programmer's original specification: any calls to `extract_words` will be calls to `profile(extract_words)` instead.

There are different implementation techniques for achieving this style of programming in different languages. A slight variation in Python is to use decorators, although that violates the third constraint of the style, as formulated here.

19.4 HISTORICAL NOTES

The idea of "advising" a function to include additional behavior to it externally was first described in a PhD thesis by Warren Teitelman in 1966. That work was done in the context of LISP. *Advice* found its way to several flavors of LISP during the 1970s. This work had a strong influence in the Aspect-Oriented Programming (AOP) style, developed in the 1990s at Xerox PARC by a group led by Gregor Kiczales, of which I was a part.

AOP is a form of restrained reflection for allowing programmers to define *aspects* of the programs. *Aspects* are concerns of the applications that tend to be scattered all over the code because they naturally affect many of its components. Typical aspects are tracing and profiling. Over the years, people have used this concept to localize in one place of their programs functionality that would be scattered otherwise.

19.5 FURTHER READING

Baldi, P., Lopes, C., Linstead, E. and Bajracharya, S. (2008). A theory of aspects as latent topics. *ACM Conference on Object-Oriented Programming, Systems, Languages and Applications (OOPSLA'08)*.
Synopsis: A more recent information-theoretic perspective on aspects.

Kiczales, G., Lamping, J., Mendhekar, A., Maeda, C., Lopes, C., Loingtier, J.-M. and Irwin, J. (1997). Aspect-oriented programming. *European Conference on Object-Oriented Programming (ECOOP'97)*.
Synopsis: The original paper for AOP co-authored by my group at Xerox PARC led by Gregor Kiczales.

Teitelman, W. (1966). PILOT: A step towards man-computer symbiosis. PhD Thesis, MIT. Available at:
ftp://publications.ai.mit.edu/ai-publications/pdf/AITR-221.pdf
Synopsis: The original idea of "advice." Chapter 3 of the thesis describes the concept.

19.6   GLOSSARY

Aspect: (1) A program concern whose implementation within the established problem decomposition defies textual localization using non-reflective composition mechanisms. (2) A topic of the source code with high entropy.

19.7   EXERCISES

19.1   *Another language.* Implement the example program in another language, but preserve the style.

19.2   *Decorate it.* Implement the profile aspect with a decorator. What do you see as pros and cons of that alternative?

19.3   *Quantification.* In the example program, we specify which functions to affect by providing a list of functions (line #40). Extend this little language by allowing the specification of "all functions in scope" in addition to specifying names; choose your syntax as you please.

19.4   *Tracing.* Add another aspect to the example program for tracing functions. That is, the following should be printed out in the beginning of functions:

```
Entering <function name>
```

and at the end of functions:

```
Exiting <function name>
```

This aspect should be in addition to the profile aspect that is already there. Functions should exhibit both the profile and the tracing aspects.

19.5   *Things.* Take the example program in the *Things* style (Chapter 11) and apply the profile aspect to the following methods: run in WordFrequencyController and the constructor of DataStorageManager.

19.6   *A different task.* Apply the profile aspect to one of the tasks proposed in the Prologue.

## CHAPTER 20

Plugins

20.1 CONSTRAINTS

▷ The problem is decomposed using some form of abstraction (procedures, functions, objects, etc.).

▷ All or some of those abstractions are physically encapsulated into their own, usually pre-compiled, packages. Main program and each of the packages are compiled independently. These packages are loaded dynamically by the main program, usually in the beginning (but not necessarily).

▷ Main program uses functions/objects from the dynamically loaded packages, without knowing which exact implementations will be used. New implementations can be used without having to adapt or recompile the main program.

▷ Existence of an external specification of which packages to load. This can be done by a configuration file, path conventions, user input or other mechanisms for external specification of code to be loaded at runtime.

20.2 A PROGRAM IN THIS STYLE

tf-19.py:

```python
1  #!/usr/bin/env python
2  import sys, configparser, importlib.machinery
3
4  def load_plugins():
5      config = configparser.ConfigParser()
6      config.read("config.ini")
7      words_plugin = config.get("Plugins", "words")
8      frequencies_plugin = config.get("Plugins", "frequencies")
9      global tfwords, tffreqs
10     tfwords = importlib.machinery.SourcelessFileLoader('tfwords',
           words_plugin).load_module()
11     tffreqs = importlib.machinery.SourcelessFileLoader('tffreqs',
           frequencies_plugin).load_module()
12
13 load_plugins()
14 word_freqs = tffreqs.top25(tfwords.extract_words(sys.argv[1]))
15
16 for (w, c) in word_freqs:
17     print(w, '-', c)
```

config.ini:

```
1  [Plugins]
2  ;; Options: plugins/words1.pyc, plugins/words2.pyc
3  words = plugins/words1.pyc
4  ;; Options: plugins/frequencies1.pyc, plugins/frequencies2.pyc
5  frequencies = plugins/frequencies1.pyc
```

words1.py:

```python
1  import sys, re, string
2
3  def extract_words(path_to_file):
4      with open(path_to_file) as f:
5          str_data = f.read()
6      pattern = re.compile('[\W_]+')
7      word_list = pattern.sub(' ', str_data).lower().split()
8
9      with open('../stop_words.txt') as f:
10         stop_words = f.read().split(',')
11     stop_words.extend(list(string.ascii_lowercase))
12
13     return [w for w in word_list if not w in stop_words]
```

words2.py:

```python
1  import sys, re, string
2
3  def extract_words(path_to_file):
4      words = re.findall('[a-z]{2,}', open(path_to_file).read().
           lower())
5      stopwords = set(open('../stop_words.txt').read().split(','))
```

```
6        return [w for w in words if w not in stopwords]
```

frequencies1.py:

```
1   import operator
2
3   def top25(word_list):
4       word_freqs = {}
5       for w in word_list:
6           if w in word_freqs:
7               word_freqs[w] += 1
8           else:
9               word_freqs[w] = 1
10      return sorted(word_freqs.items(), key=operator.itemgetter(1),
        reverse=True)[:25]
```

frequencies2.py

```
1   import operator, collections
2
3   def top25(word_list):
4       counts = collections.Counter(w for w in word_list)
5       return counts.most_common(25)
```

20.3 COMMENTARY

THIS STYLE is at the center of software evolution and customization. Developing software that is meant to be extended by others, or even by the same developers but at a later point in time, carries a set of challenges that don't exist in close-ended software.

Let's look at the example program. The main idea is to devoid the main program of important implementation details, leaving it only as a "shell" that executes two term frequency functions. In this case, we partition the term frequency application into two separate steps: in the first step, called extract_words, we read the input file and produce a list of non-stop words; in the second step, called top25, we take that list of words, count their occurrences, and return the 25 most frequently occurring words and their counts. These two steps can be seen in line #14 of tf-19.py. Note that the main program, tf-19.py, has no knowledge of the functions tfwords.extract_words and tffreqs.top25, other that they exist, hopefully. We want to be able to choose the implementation of those functions at some later point in time, maybe even allow the users of this program to provide their own implementations.

By the time it runs, the main program needs to know which functions to use. That is specified externally in a configuration file. As such, the first thing that our main program does before calling the term frequency functions, is to load the corresponding plugins (line #13). load_plugins (lines #4–11) starts by reading the configuration file config.ini (lines #5–6) and extracting the settings for the two functions (lines #7–8). It is assumed that the configuration file contains a section named Plugins where two configuration variables can be found: words (line #7) and frequencies (line #8). The value of those variables is supposed to be paths to pre-compiled Python code.

Before explaining the next 3 lines, let's take a look at what the configuration file looks like – see config.ini. We are using a well-known configuration format known as INI, which has pervasive support in the programming world. Python's standard library supports it via the ConfigParser module. INI files are very simple, consisting of one or more sections denoted by [SectionName] in a single line, under which configuration variables and their values are listed as key-value pairs (name=value), one per line. In our case, we have one section called [Plugins] (line #1) and two variables: words (line #3) and frequencies (line #5). Both of the variables hold values that are meant to be paths in the file system; for example, words is set to plugins/words1.pyc, meaning that we want to use a file that is in a sub-directory of the current directory. We can change which plugin to use by changing the values of these variables.

Coming back to tf-19.py, lines #5–8 are, then, reading this configuration file. The next 3 lines (#9–11) deal with loading the code dynamically from the files that we have specified in the configuration. We do this using

Python's imp module, which provides a reflective interface to the internals of the import statement. In line #9, we declare two global variables, tfwords and tffreqs, that are meant to be modules themselves. Then in lines #10 and #11 we load the code found in the specified paths and bind it to our module variables. imp.load_compiled takes a name and a path to a pre-compiled Python file and loads that code into memory, returning the compiled module object – we then need to bind that object to our module name, so that we can use it in the rest of the main program (specifically in line #14).

The rest of the example program – words1.py, words2.py, frequencies1.py and frequencies2.py – shows the different implementations of our term frequency functions. words1.py and words2.py provide alternatives for the extract_words function; frequencies1.py and frequencies2.py provide alternatives for the top25 function.[1]

20.4   THIS STYLE IN SYSTEMS DESIGN

It is important to understand the alternatives to this style that achieve the same goal of supporting different implementations of the same functions. It is also important to understand those alternatives' limits, and the benefits of this style of programming.

When wanting to support different implementations of the same functions, one can protect the callers of those functions using well-known design patterns such as the Factory pattern: the callers request a specific implementation, and a Factory method returns the right object. In their simplest form, Factories are glorified conditional statements over a number of pre-defined alternatives. Indeed, the simplest way to support alternatives is with conditional statements.

Conditional statements and related mechanisms assume that the set of alternatives is known at program design time. When that is the case, the Plugins style is an overkill, and a simple Factory pattern will serve the goal well. However, when the set of alternatives is open-ended, using conditionals quickly becomes a burden: for every new alternative, we need to edit the Factory code and compile it again. Furthermore, when the set of alternatives is meant to be open to third parties who don't necessarily have access to the source code of the base program, it is simply not possible to achieve the goal using hardcoded alternatives; dynamic code loading becomes necessary. Modern frameworks have embraced this style of programming for supporting usage-sensitive customizations.

Modern operating systems also support this style via shared, dynamically linked libraries (e.g. .so in Lunix and .DLL in Windows).

However, when abused, software written in this style can become a "configuration hell," with dozens of customization points, each with many different

---

[1]Note that in order for our program to work as is, these files need to be compiled first into .pyc files.

alternatives that can be hard to understand. Furthermore, when alternatives for different customization points have dependencies among themselves, software may fail mysteriously, because the simple configuration languages in use today don't provide good support for expression of dependencies between external modules.

20.5 HISTORICAL NOTES

The origins of this style are somewhat foggy, but seem to spread across two separate lines of work: distributed systems architecture and the need to extend standalone applications with third-party code.

Mesa, a programming language designed at Xerox PARC in the 1970s and used in the Xerox Star office automation system, included a configuration language that was used to inform the linker how to bind together a set of modules into a complete system. C/Mesa featured separate interface and implementation modules (similar to Abstract Things), so a C/Mesa program could wire together the exports and imports of implementation modules. This was used to assemble together different variants of the operating system.

By the mid-1980s, several sophisticated networked control systems were being built that required careful thinking about the system as a collection of independent components that needed to be connected, and that could potentially be replaced by other components. As such, configuration languages started to be proposed. These configuration languages embodied the concept of separating the functional components from their interconnections, suggesting "configuration programming" as a separate concern. This line of work continued through the 1990s under what is now known as *software architecture*, and configuration languages became Architecture Description Languages (ADLs). Many ADLs proposed during the 1990s, although powerful, were simply languages for the analysis of systems, and were not executable. This was due, in part, to the fact that linking components at runtime was a hard thing to do with the mainstream programming language technology of the time, which was heavily C-based. The ADLs that were executable used niche languages that were not mainstream.

During the 1990s, several desktop applications already supported plugins. For example, PhotoShop had that concept from very early on, as it enabled a clean separation of the "core" application from the several image filters that could be added, possibly by end-users; it also allowed customizations of image processing functions on the hardware of the desktops.

The advent of mainstream programming languages with reflective capabilities changed the landscape of this work, as it suddenly became possible, and trivially easy, to link components at runtime. Java frameworks, such as Spring, were the first to embrace the new capabilities brought by reflection. As many more languages started to embrace reflection, this style of engineering systems became commonplace in industry under the names "dependency injection" and "plugins." Within these practices, ADLs are back to being simple declarative configuration languages, such as INI or XML.

20.6 FURTHER READING

Fowler, M. (2004). Inversion of control containers and the dependency injection pattern. Blog post available at:
http://www.martinfowler.com/articles/injection.html
Synopsis: Martin Fowler explains inversion of control and dependency injection in the context of OOP frameworks.

Kramer, J., Magee, J., Sloman, M. and Lister, A. (1983). CONIC: An integrated approach to distributed computer control systems. *IEE Proceedings* 130(1): 1–10.
Synopsis: The description of one of the first Architecture Description Languages (ADL) to be called as such.

Mitchell, J., Maybury, W. and Sweet, R. (1979). Mesa Language Manual. Xerox PARC Technical Report CSL-79-3. Available at:
http://bitsavers.trailing-edge.com/pdf/xerox/mesa/5.0_1979/
documentation/CSL_79-3_Mesa_Language_Manual_Version_5.0_Apr79.pdf
Synopsis: Mesa was a really interesting language. It was a Modula-like language, so very focused on modularity issues. Mesa programs consisted of definition files specifying interfaces plus one or more program files specifying the implementation of the procedures in the interfaces. Mesa was a major influence on the design of other languages, such as Modula-2 and Java.

20.7 GLOSSARY

Third-party development: Development for a piece of software done by a different group of developers than those developing that software. Third-party development usually involves having access only to the binary form of the software, not its source code.

Dependency injection: A collection of techniques that support importing function/object implementations dynamically.

Plugin: (*aka* Addon) A software component that adds a specific set of behaviors into an executing application, without the need for recompilation.

20.8 EXERCISES

20.1 *Another language.* Implement the example program in another language, but preserve the style.

20.2 *Different extraction.* Provide a third alternative to `extract_words`.

20.3 *Close-ended.* Suppose that words1.py, words2.py, frequencies1.py and frequencies2.py are the only possible alternatives to ever be considered in the example program. Show how you would transform the program **away** from the Plugins style.

20.4 *Print out alternatives.* The example program hardcodes the printout of the word frequencies at the end (lines #16–17). Transform that into the Plugins style and provide at least two alternatives for printing out the information at the end.

20.5 *Link source code.* Modify the load_plugins function so that it can also load modules with Python source code.

20.6 *A different task.* Write one of the tasks proposed in the Prologue using this style.

VI

Adversity

IV

Advaita

When programs execute, abnormal things may happen, either intentionally (by malicious attacks) or unintentionally (by programmer's overlook or unexpected failures in hardware). Dealing with them is perhaps one of the most complicated activities in program design. One approach to dealing with abnormalities is to be *oblivious* to them. This can be done by either (1) assuming that errors don't occur or (2) not caring if they occur. For the purpose of focusing on specific constraints without distractions, obliviousness is the style followed in this book – except in the next five styles. The next five chapters – *Constructivist, Tantrum, Passive Aggressive, Declared Intentions* and *Quarantine* – reflect five different approaches to dealing with adversity in programs. They are all instances of a more general style of programming known as *defensive programming*, which is very much the opposite of the *oblivious* style. A comparative analysis of the first three variations of defensive programming is presented at the end of Chapter 23.

## CHAPTER 21

Constructivist

21.1 CONSTRAINTS

▷ Every single function checks the sanity of its arguments and either returns something sensible when the arguments are unreasonable or assigns them reasonable values.

▷ All code blocks check for possible errors and escape the block when things go wrong, setting the state to something reasonable, and continuing to execute the rest of the function.

21.2 A PROGRAM IN THIS STYLE

```
1  #!/usr/bin/env python
2  import sys, re, operator, string, inspect
3
4  #
5  # The functions
6  #
7  def extract_words(path_to_file):
8      if type(path_to_file) is not str or not path_to_file:
9          return []
10
11     try:
12         with open(path_to_file) as f:
13             str_data = f.read()
14     except IOError as e:
15         print("I/O error({0}) when opening {1}: {2}".format(e.
               errno, path_to_file, e.strerror))
16         return []
17
18     pattern = re.compile('[\W_]+')
19     word_list = pattern.sub(' ', str_data).lower().split()
20     return word_list
21
22 def remove_stop_words(word_list):
23     if type(word_list) is not list:
24         return []
25
26     try:
27         with open('../stop_words.txt') as f:
28             stop_words = f.read().split(',')
29     except IOError as e:
30         print("I/O error({0}) when opening ../stops_words.txt: {1}
               ".format(e.errno, e.strerror))
31         return word_list
32
33     stop_words.extend(list(string.ascii_lowercase))
34     return [w for w in word_list if not w in stop_words]
35
36 def frequencies(word_list):
37     if type(word_list) is not list or word_list == []:
38         return {}
39
40     word_freqs = {}
41     for w in word_list:
42         if w in word_freqs:
43             word_freqs[w] += 1
44         else:
45             word_freqs[w] = 1
46     return word_freqs
47
48 def sort(word_freq):
49     if type(word_freq) is not dict or word_freq == {}:
50         return []
51
```

```
52      return sorted(word_freq.items(), key=operator.itemgetter(1),
            reverse=True)
53
54  #
55  # The main function
56  #
57  filename = sys.argv[1] if len(sys.argv) > 1 else "../input.txt"
58  word_freqs = sort(frequencies(remove_stop_words(extract_words(
        filename))))
59
60  for tf in word_freqs[0:25]:
61      print(tf[0], '-', tf[1])
```

21.3 COMMENTARY

I N THIS STYLE, programs are mindful of possible abnormalities; they don't ignore them, but they take a constructivist approach to the problem: they incorporate practical heuristics in order to fix the problems in the service of getting the job done. They defend the code against possible errors of callers and providers by using reasonable fallback values whenever possible so that the program can continue.

Let's look at the example program, starting from the bottom. In all previous examples, we are not checking whether the user gave a file name in the command line – in true oblivious style, we assume that the file name argument will be there, and if it's not, the program crashes. In this program, we are now checking whether the user has given a file name (line #57) and if they didn't, our program falls back to computing the term frequency of an existing test file, input.txt.

A similar approach can be seen in other parts of this program. For example, in the function extract_words, lines #11–16, when there are errors opening or reading the given file name, the function simply acknowledges that and returns an empty list of words, allowing the program to continue based on that empty list of words. And in the function remove_stop_words, lines #26–31, if there are errors regarding the file that contains the stop words, that function simply echoes back the word list that it received, effectively not filtering for stop words.

The *Constructivist* style of dealing with the inconveniences of errors can have a tremendous positive effect on user experience. However, it comes with some perils that need to be carefully considered.

First, when the program assumes some fallback behavior without notifying the user, the results may be puzzling. For example, running this program without the file name:

```
$ python tf-21.py
mostly  -  2
live  -  2
africa  -  1
tigers  -  1
india  -  1
lions  -  1
wild  -  1
white  -  1
```

This produces a result that the user may not understand. Where did those words come from? In assuming fallback values, it is important to let the user know what is going on.

The program behaves better when the file doesn't exist:

```
$ python tf-21.py
I/O error(2) when opening foo: No such file or directory
```

Even though the functions continue to execute on empty lists, the user is made aware that something didn't quite work as expected.

The second peril has to do with the heuristics used for fallback strategies. Some of them may be more confusing than an explicit error, or even misleading. For example, if in lines #11–16, upon encountering a file (provided by the user) that doesn't actually exist, we would fall back to opening `input.txt`, the user would be misled to thinking that the file that they provided had the resulting term frequencies. Clearly this is false. At the very least, if we would decide on that fallback strategy, we would need to warn the user about the situation ("That file doesn't exist, but here are the results for another one").

21.4   THIS STYLE IN SYSTEMS DESIGN

Many popular computer languages and systems take this approach to adversity. The rendering of HTML pages in Web browsers, for example, is notorious for being constructivist: even if the page has syntax errors, or inconsistencies, the browser will try to render it as best as possible. Python itself also takes this approach in many situations, such as when obtaining ranges of lists beyond their length (see **Bounds** in page xx).

Modern user-facing software also tends to take this approach, sometimes with the use of heavy heuristic machinery underneath. When entering keywords in search engines, the search engines often correct spelling mistakes and present results for the correctly spelled words, instead of taking the user input literally.

Trying to guess the intention behind an input error is a very nice thing to do, as long as the system is in a position to guess right most of the time. People tend to lose trust in systems that make wrong guesses.

21.5   EXERCISES

21.1   *Another language.* Implement the example program in another language, but preserve the style.

21.2   *A different task.* Write one of the tasks proposed in the Prologue using this style.

## CHAPTER 22

Tantrum

22.1 CONSTRAINTS

▷ Every single procedure and function checks the sanity of its arguments and refuses to continue when the arguments are unreasonable.

▷ All code blocks check for all possible errors, possibly log context-specific messages when errors occur, and pass the errors up the function call chain.

22.2   A PROGRAM IN THIS STYLE

```python
1  #!/usr/bin/env python
2
3  import sys, re, operator, string, traceback
4
5  #
6  # The functions
7  #
8  def extract_words(path_to_file):
9      assert(type(path_to_file) is str), "I need a string!"
10     assert(path_to_file), "I need a non-empty string!"
11
12     try:
13         with open(path_to_file) as f:
14             str_data = f.read()
15     except IOError as e:
16         print("I/O error({0}) when opening {1}: {2}! I quit!".
                 format(e.errno, path_to_file, e.strerror))
17         raise e
18
19     pattern = re.compile('[\W_]+')
20     word_list = pattern.sub(' ', str_data).lower().split()
21     return word_list
22
23 def remove_stop_words(word_list):
24     assert(type(word_list) is list), "I need a list!"
25
26     try:
27         with open('../stop_words.txt') as f:
28             stop_words = f.read().split(',')
29     except IOError as e:
30         print("I/O error({0}) when opening ../stops_words.txt:
                 {1}! I quit!".format(e.errno, e.strerror))
31         raise e
32
33     stop_words.extend(list(string.ascii_lowercase))
34     return [w for w in word_list if not w in stop_words]
35
36 def frequencies(word_list):
37     assert(type(word_list) is list), "I need a list!"
38     assert(word_list != []), "I need a non-empty list!"
39
40     word_freqs = {}
41     for w in word_list:
42         if w in word_freqs:
43             word_freqs[w] += 1
44         else:
45             word_freqs[w] = 1
46     return word_freqs
47
48 def sort(word_freq):
49     assert(type(word_freq) is dict), "I need a dictionary!"
50     assert(word_freq != {}), "I need a non-empty dictionary!"
51
52     try:
```

```
53        return sorted(word_freq.items(), key=operator.itemgetter
              (1), reverse=True)
54    except Exception as e:
55        print("Sorted threw {0}".format(e))
56        raise e
57
58 #
59 # The main function
60 #
61 try:
62    assert(len(sys.argv) > 1), "You idiot! I need an input file!"
63    word_freqs = sort(frequencies(remove_stop_words(extract_words(
              sys.argv[1]))))
64
65    assert(type(word_freqs) is list), "OMG! This is not a list!"
66    assert(len(word_freqs) > 25), "SRSLY? Less than 25 words!"
67    for (w, c) in word_freqs[0:25]:
68        print(w, '-', c)
69 except Exception as e:
70    print("Something wrong: {0}".format(e))
71    traceback.print_exc()
```

22.3 COMMENTARY

THIS STYLE is as defensive as the previous one: the same possible errors are being checked. But the way it reacts when abnormalities are detected is quite different: the functions simply refuse to continue.

Let's look at the example program, again starting at the bottom. In line #62, we are not just checking that there is a file name given in the command line, but we are *asserting* that it must exist, or else it throws an exception – the assert function throws the AssertionError exception when the stated condition is not met.

A similar approach can be seen in other parts of the program. In the function extract_words, lines #9 and #10, we are asserting that the argument meets certain conditions, or else the function throws an exception. In lines #12–17, if the opening or reading of the file throws an exception, we are catching it right there, printing a message about it, and passing the exception up the stack for further catching. Similar code – i.e. assertions, and local exception handling – can be seen in all the other functions.

Stopping the program's execution flow when abnormalities happen is one way to ensure that those abnormalities don't cause damage. In many cases, it may be the only option, as fallback strategies may not always be good or desirable.

This style has one thing in common with the *Constructivist* style of the previous chapter: it is checking for errors, and handling them, in the local context in which the errors may occur. The difference here is that the fallback strategies of the *Constructivist* style are interesting parts of the program in themselves, whereas the cleanup and exit code of the *Tantrum* style is not.

This kind of local error checking is particularly visible in programs written in languages that don't have exceptions. C is one of those languages. When guarding against problems, C programs check locally whether errors have occurred, and, if so, either use reasonable fallback values (*Constructivist*) or escape the function in the style explained here. In languages without exception handling, like C, the abnormal return from functions is usually flagged using error codes in the form of negative integers, null pointers, or global variables (e.g. errno), which are then checked in the call sites.

Dealing with abnormalities in this way can result in verbose boilerplate code that distracts the reader from the actual goals of the functions. It is quite common to encounter portions of the programs written in this style with one line of functional code followed by a long sequence of conditional blocks that check for the occurrence of various errors, each one returning an error at the end of the block.

In order to avoid some of the verbosity of this style, advanced C programmers sometimes resort to using C's GOTO statement. One of the main advantages of GOTOs is the fact that they allow non-local escapes, avoiding boilerplate, distracting code when dealing with errors, while supporting a single exit point out of functions. GOTOs allow us to express our displeasure with errors in a more contained, succinct form. But GOTOs have long been discouraged, or outright banned, from mainstream programming languages, for all sorts of good reasons.

22.4 THIS STYLE IN SYSTEMS DESIGN

Computers are dumb machines that need to be told exactly and unambiguously what to do. Computer software inherited that trait. Many software systems don't make much effort in trying to guess the intentions behind wrong inputs (from users or other components); it is much easier and risk-free to simply refuse to continue. Therefore this style is seen pervasively in software. Worse, many times the errors are flagged with incomprehensible error messages that don't inform the offending party in any actionable way.

When being pessimistic about adversity, it is important to at least let the other party know what was expected and why the function/component is refusing to continue.

22.5 FURTHER READING

IBM (1957). The FORTRAN automatic coding system for the IBM 704 EDPM. Available at:
http://www.softwarepreservation.org/projects/FORTRAN/manual/
Prelim_Oper_Man-1957_04_07.pdf
Synopsis: The original FORTRAN manual, showing a long list of possible error codes and what to do with them. The list mixes machine (hardware) errors with human (software) errors. Some of the human errors are syntactic while others are a bit more interesting. For example, error 430 is described as "Program too complex. Simplify or do in 2 parts (too many basic blocks)."

22.6 GLOSSARY

Error code: Enumerated messages that denote faults in specific components.

22.7 EXERCISES

22.1 *Another language.* Implement the example program in another language, but preserve the style.

22.2 *A different task.* Write one of the tasks proposed in the Prologue using this style.

## CHAPTER 23

Passive Aggressive

23.1 CONSTRAINTS

▷ Every single procedure and function checks the sanity of its arguments and refuses to continue when the arguments are unreasonable, jumping out of the function.

▷ When calling out other functions, program functions only check for errors if they are in a position to react meaningfully.

▷ Exception handling occurs at higher levels of function call chains, wherever it is meaningful to do so.

23.2 A PROGRAM IN THIS STYLE

```python
#!/usr/bin/env python
import sys, re, operator, string

#
# The functions
#
def extract_words(path_to_file):
    assert(type(path_to_file) is str), "I need a string! I quit!"
    assert(path_to_file), "I need a non-empty string! I quit!"

    with open(path_to_file) as f:
        data = f.read()
    pattern = re.compile('[\W_]+')
    word_list = pattern.sub(' ', data).lower().split()
    return word_list

def remove_stop_words(word_list):
    assert(type(word_list) is list), "I need a list! I quit!"

    with open('../stop_words.txt') as f:
        stop_words = f.read().split(',')
    # add single-letter words
    stop_words.extend(list(string.ascii_lowercase))
    return [w for w in word_list if not w in stop_words]

def frequencies(word_list):
    assert(type(word_list) is list), "I need a list! I quit!"
    assert(word_list != []), "I need a non-empty list! I quit!"

    word_freqs = {}
    for w in word_list:
        if w in word_freqs:
            word_freqs[w] += 1
        else:
            word_freqs[w] = 1
    return word_freqs

def sort(word_freqs):
    assert(type(word_freqs) is dict), "I need a dictionary! I quit!"
    assert(word_freqs != {}), "I need a non-empty dictionary! I quit!"

    return sorted(word_freqs.items(), key=operator.itemgetter(1),
        reverse=True)

#
# The main function
#
try:
    assert(len(sys.argv) > 1), "You idiot! I need an input file! I quit!"
    word_freqs = sort(frequencies(remove_stop_words(extract_words(
        sys.argv[1]))))
```

```
50
51      assert(len(word_freqs) > 25), "OMG! Less than 25 words! I QUIT
        !"
52      for tf in word_freqs[0:25]:
53          print(tf[0], '-', tf[1])
54  except Exception as e:
55          print("Something wrong: {0}".format(e))
```

23.3   COMMENTARY

JUST LIKE THE PREVIOUS STYLE, this style deals with caller mistakes (pre-conditions) and general execution errors by skipping the rest of the execution in a call chain. However, it does it differently from the *Tantrum* style: rather than scattering error handling code all over the program, as if throwing a very vocal tantrum, error handling is contained in just one place. But the result is still the same: any functions down the call chain aren't executed. Such is the Passive Aggressive behavior in the face of adversity.

Let's look at the example program. Like the Tantrum style, the program's functions check for the validity of input arguments, returning an error immediately if they aren't valid – see assertions in lines #8, #9, #18, #27, #28, #39, #40, #48, and #51. Unlike the Tantrum style, the possible errors resulting from calls to other functions, such as library functions, aren't explicitly handled at the points at which they are called. For example, the opening and reading of the input file in lines #11–12 isn't guarded by a try-except clause; if an exception occurs there, it will simply break the execution of that function and pass the exception up the call chain until it reaches some exception handler. In our case, that handler exists at the top-most level, in lines #54–55.

Certain programming languages are, by design, hostile to supporting the Passive Aggressive style, encouraging either the Constructivist or the Tantrum styles. C is a good example of such a language. But it is technically possible to use this style in languages that don't support exceptions as we have come to know them in mainstream programming languages. Two examples: (1) Haskell supports this style via the Exception monad, and without any special language support for exceptions; and (2) many experienced C programmers have come to embrace the use of GOTO for better modularization of error handling code, which results in a more Passive Aggressive attitude to error handling.

23.4   HISTORICAL NOTES

Exceptions were first introduced in PL/I in the mid-1960s, although their use there was a bit controversial. For example, reaching the end of the file was considered an exception. In the early 1970s, LISP also had exception handling.

23.5   FURTHER READING

Abrahams, P. (1978). The PL/I Programming Language. Courant Mathematics and Computing Laboratory, New York University. Available at: http://www.iron-spring.com/abrahams.pdf
Synopsis: The PL/I specification. PL/I was the first language supporting some version of exceptions.

23.6   GLOSSARY

Exception: A situation outside the normal expectations in the program execution.

23.7   EXERCISES

23.1   *Another language.* Implement the example program in another language, but preserve the style.

23.2   *Abnormalities.* Make abnormalities occur for this program, both the program as a whole and the individual functions. Show how the program behaves in the face of those abnormalities. Tip: write test cases that test for situations that will make the program fail.

23.3   *The exception master object.* Write a version of the term-frequency program that emulates exceptions using a "master object" similar to that seen in Chapter 10. For that, there should be no try-catch block in the main function. Instead, the master object should catch exceptions. The master object's role is to unfold the computation and at every step, check if there were errors; if so, no further functions should be called. Test it, for example, by giving an erroneous name to the stop words file. You can either start with this chapter's example code or with the code in Chapter 10 (or your version of it in another language). The main function of your resulting program should use *bind* to chain functions or objects.

23.4   *A different task.* Write one of the tasks proposed in the Prologue using this style.

23.8 CONSTRUCTIVIST VS. TANTRUM
VS. PASSIVE AGGRESSIVE

These three styles – Constructivist, Tantrum and Passive Aggressive – reflect three different approaches to dealing with adversity.

Exceptions were introduced as a structured, well-behaved, restrained alternative to GOTOs for the specific purpose of dealing with abnormalities. Exceptions don't allow us to jump to arbitrary places of the program, but they allow us to return to arbitrary functions in the call stack, avoiding unnecessary boilerplate code. Exceptions are a more contained form of protesting against obstacles in our way. They are the image of Passive Aggressive behavior ("I'm not protesting now, but this is not right and I'll protest eventually").

But even when languages support exceptions, not all programs written in those languages are passive aggressive with respect to abnormalities, as demonstrated here. Two factors may play a role in this.

Often, the first instinct of relatively inexperienced programmers who start learning about exceptions is to use the Tantrum style, because they aren't comfortable about letting the error go without checking it locally where it first occurs. It takes some time to gain confidence in the exception mechanism. In other cases, it's the programming language that encourages tantrums. Java, for example, imposes statically checked exceptions; this forces programmers to have to declare those exceptions in the method signatures when they simply wish to ignore them. Given that declaring exceptions in method signatures can quickly become a time-consuming burden, it is often simpler to catch the exceptions right where they may occur, resulting in code with exception "tantrums." It is not unusual to see Java programs that use C-style tantrums by catching exceptions locally and returning error codes instead.

In general, when we decide to deal with abnormalities, the Passive Aggressive style is preferred over the Tantrum style. One should not catch an exception (that is to say "protest") prematurely, when it's not clear how to recover from it; we also shouldn't do it just to log that it happened – the call stack is part of the exception information, wherever it is caught. Often, it's the caller of our function, or even higher above, who has the right context for dealing with the problem, so, unless there is some meaningful local processing to be done when an abnormality happens, it's better to let the exception go up the call chain.

In many applications, though, the Constructivist style has several advantages over the other two. By assuming reasonable fallback values to erroneous function arguments and returning reasonable fallback values when things go wrong within a function, we allow the program to continue, and do its best at the task that it is supposed to do.

## CHAPTER 24

Declared Intentions

24.1 CONSTRAINTS

▷ Existence of a type enforcer.

▷ Procedures and functions declare what types of arguments they expect.

▷ If callers send arguments of types that aren't expected, type errors are raised, and the procedures/functions are not executed.

24.2   A PROGRAM IN THIS STYLE

```python
#!/usr/bin/env python
import sys, re, operator, string, inspect

#
# Decorator for enforcing types of arguments in method calls
#
class AcceptTypes():
    def __init__(self, *args):
        self._args = args

    def __call__(self, f):
        def wrapped_f(*args):
            for i in range(len(self._args)):
                if type(args[i]) != self._args[i]:
                    raise TypeError("Expecting %s got %s" % (str(
                        self._args[i]), str(type(args[i]))))
            return f(*args)
        return wrapped_f
#
# The functions
#
@AcceptTypes(str)
def extract_words(path_to_file):
    with open(path_to_file) as f:
        str_data = f.read()
    pattern = re.compile('[\W_]+')
    word_list = pattern.sub(' ', str_data).lower().split()
    with open('../stop_words.txt') as f:
        stop_words = f.read().split(',')
    stop_words.extend(list(string.ascii_lowercase))
    return [w for w in word_list if not w in stop_words]

@AcceptTypes(list)
def frequencies(word_list):
    word_freqs = {}
    for w in word_list:
        if w in word_freqs:
            word_freqs[w] += 1
        else:
            word_freqs[w] = 1
    return word_freqs

@AcceptTypes(dict)
def sort(word_freq):
    return sorted(word_freq.items(), key=operator.itemgetter(1),
        reverse=True)

word_freqs = sort(frequencies(extract_words(sys.argv[1])))
for (w, c) in word_freqs[0:25]:
    print(w, '-', c)
```

24.3   COMMENTARY

THERE IS A CATEGORY of programming abnormalities that, from very early on in the history of computers, has been known to be problematic: type mismatches. That is, a function expects an argument of a certain type, but is given a value of another type; or a function returns a value of a certain type which is then used by the caller of the function as if it were a value of another type. This is problematic because values of different types usually have different memory sizes, which means that when type mismatches occur, memory can be overwritten and made inconsistent.

Luckily, these abnormalities are relatively easy to deal with – at least in comparison to all other abnormalities that can happen during program execution – and the issue has been largely solved for quite some time in mainstream programming languages by means of *type systems*. All modern high-level programming languages have a type system,[1] and data types are checked in various points of program development and execution.

Python, too, has a type system, and a very strong one. For example, we can index these values:

```
>>> ['F', 'a', 'l', 's', 'e'][3]
's'
>>> "False"[3]
's'
```

but we get a type error if we try to index this other value:

```
>>> False[3]
Traceback (most recent call last):
  File "<stdin>", line 1, in <module>
TypeError: 'bool' object has no attribute '__getitem__'
```

meaning that the Python interpreter is not fooled by our attempt at indexing the letter "s" of the Boolean value False, and refuses to do what we ask by throwing an exception.

Python performs type checks dynamically, meaning that only when the program runs do values get type checked. Other languages perform type checking ahead of time – those are called static type checking languages. Java and Haskell are examples of statically type checked programming languages. Java and Haskell are also good examples of opposite approaches to static type checking: while Java requires programmers to declare variable types explicitly, Haskell supports implicit type declarations thanks to its type inference capabilities.

Although never proved empirically, many people believe that knowing the types of values ahead of time, rather than waiting for the runtime to throw an error, is a good software engineering practice, especially in the development

---

[1]Some type systems enforce types more than others.

of large, multi-person projects. This belief is the basis for the style presented in this chapter, *Declared Intentions*.

Let's look at the example program. The program uses functional abstraction like many other examples seen before, defining 3 main functions: extract_words (lines #22–30), frequencies (lines #33–40) and sort (lines 43–44). We know that these functions will only work well if the arguments that are passed to them are of certain types. For example, the function extract_words will not work if the caller passes, say, a list. So, instead of hiding that knowledge, we can expose it to the callers.

That is exactly what the example program does via the declarations *@AcceptTypes(...)* just above each of the function definitions – see lines #21, #32 and #42. These declarations are implemented with a Python *decorator*. A decorator is another reflective feature of Python that allows us to change functions, methods or classes without changing their source code. A new decorator instance is created every time a decorator is used.[2] Let's take a closer look at this decorator.

The AcceptTypes decorator's constructor (lines #8–9) executes in every usage. The constructor takes a list of arguments – our *string* in line #21, *list* in line #32 and *dict* in line #42 – and simply stores them. When the decorated functions are called, the decorator's __call__ method (lines #11–17) is called first. In our case, we check whether the types of the provided arguments to the functions are the same as the ones that had been declared at the time of the functions' declaration (lines #13–14). If they aren't, a type error is raised. This way, we guarantee that our functions are not executed when their arguments don't match the expectation. But, more importantly, *we have stated our intentions to the callers of our functions*.

At this point, we should pause over three pertinent questions about this style and its illustration in the example program.

What is the difference between these type annotations and the existing Python type system? Our type annotations narrow down the accepted types much more than the existing type system. Take, for example, the frequencies function (lines #33–40). The argument word_list is used in line #35 as an iterable value. There are many kinds of iterable values in Python: lists, dictionaries, tuples and even strings are examples of built-in types that are iterable. Our type annotation in line #32 is saying that we expect only a list and nothing else. Python's approach to types is largely *duck typing*: a value is what a value does (if it walks like a duck, swims like a duck and quacks like a duck, it's a duck). Our type annotations are *nominal typing*: the names of the types are the foundation for type checking.

Are these type declarations the same as static typing? No. Type checking in our decorator is being done at runtime rather than before runtime. In that respect, our decorator is not giving us much more than Python's approach

---

[2]We had a previous encounter with this language feature in one of the exercises in Chapter 19, "Aspects." In fact, type declarations can be seen as an aspect of the program.

to type checking. Due to Python's approach to types, it is quite difficult to implement static type checking, although Python 3.x is coming closer to it via the new feature of function annotations. What our type annotations do, though, is make the expectations about the argument types *explicit* rather than *implicit*. It serves as documentation and warning for the callers of these functions. That is the core of this style.

What is the difference between the Declared Intentions style and the previous two styles? The *Declared Intentions* style applies only to one category óf abnormalities: type mismatches. The previous two example programs check for more than just types; for example, they check whether the given arguments are empty or have certain sizes. Those kinds of conditions are cumbersome, although not impossible, to express in terms of types.

24.4 HISTORICAL NOTES

Types in programming languages have had a long evolution that is still unfolding today. In the beginning of computers, data values had only one single numerical type, and it was entirely up to the programmers to make sure that the operations on those values made sense. In 1954, FORTRAN designers introduced a distinction between integers and floating-point numbers; that distinction was denoted by the first letter of the variables' names. This seemingly simple decision turned out to have a huge impact in the evolution of programming languages.

A few years later, Algol 60 took that distinction one step further by introducing identifier declarations for integers, reals, and Booleans. Beyond the simple integer vs. floating point distinction in FORTRAN, Algol 60 was the first major language to support compile-time type checking. During the 1960s, many languages expanded on Algol 60's type concept. Languages like PL/I, Pascal and Simula made significant contributions to the evolution of types in programming languages.

By the end of the 1960s, it was clear that static type systems were gaining solid ground in programming languages: Algol 68 had a type system so complex (including procedures as first-class values, a large variety of primitive types, type constructors, equivalence rules and coercion rules) that many found it unusable. Algol went on to influence the design of almost all the major programming languages that came after it. Static type checking was carried along with that influence.

In parallel with that line of work, and during that same time, LISP started with a very simple type system consisting only of lists and some primitive data types. This simplicity came from the theoretical work on which LISP was based – the lambda calculus. Over the years, the type system became more complex, but the foundation was unchanged: values have types; variables don't. This is the foundation for dynamic typing.

In the late 1960s, Simula, the first object-oriented language, expanded the concept of type to include classes. Instances of classes could be assigned to class-valued variables. The interface provided by these class types consisted of their declared procedures and data. All subsequent OOP languages built on this concept.

In the 1970s, work in the functional programming language ML, influenced by the *typed* version of the lambda calculus, lead to a family of type systems that are able to statically infer the types of expressions without requiring explicit type annotations. Haskell falls in this category.

As described in Chapter 20, Mesa, a language designed with physical modularization in mind, introduced typed interfaces separately from the implementation of modules. We see that concept now in Java and C#, for example.

The work on type systems is far from over. Some researchers believe that all kinds of adversity in programming can be addressed with advanced static type systems. That belief is a strong incentive to continue to devise new ways of using types. Recent work also includes optional static type checking that can be turned on and off.

24.5 FURTHER READING

Cardelli, L. (2004). Type systems. *CR Handbook of Computer Science and Engineering* 2nd ed. Ch 97. CRC Press, Boca Raton, FL.
Synopsis: One of the best overviews of types and type systems in programming languages.

Hanenberg, S. (2010). An experiment about static and dynamic type systems. *ACM Conference on Object-Oriented Programming, Systems, Languages and Applications (OOPSLA'10)*.
Synopsis: Much has been said over the years in the war between static and dynamic type checking. To date, there is no strong empirical evidence that one is better than the other. Those discussions tend to revolve around folklore and personal preferences. This is one of the few studies so far that tries to find scientific evidence one way or another.

24.6 GLOSSARY

Dynamic type checking: Type enforcement that is done during execution of the programs.

Explicit types: Type declarations that are part of the language syntax.

Implicit types: Types that do not have a presence in the language syntax.

Static type checking: Type enforcement that is done before execution of the programs.

Type coercion: The transformation of a data value from one type to another.

Type inference: The process of finding the type of an expression automatically from the leaf nodes of the expression.

Type safety: The assurance that programs will not perform instructions with type mismatches that go undetected.

24.7 EXERCISES

24.1 *Another language.* Implement the example program in another language, but preserve the style.

24.2 *Return.* The AcceptTypes decorator of the example program works only for the input arguments of functions. Write another decorator called ReturnTypes that does a similar thing for the return value(s) of functions. An example of its use is as follows:

```
@ReturnTypes(list)
@AcceptTypes(str)
def extract_words(path_to_file):
    . . .
```

24.3 *Static type checking.* Using Python 3.x, propose a mechanism for performing static type checking. Use that mechanism in some version of the example program. Hint: use function annotations.

24.4 *A different task.* Write one of the tasks proposed in the Prologue using this style.

## CHAPTER 25

Quarantine

25.1 CONSTRAINTS

▷ Core program functions have no side effects of any kind, including IO.

▷ All IO actions must be contained in computation sequences that are clearly separated from the pure functions.

▷ All sequences that have IO must be called from the main program.

25.2   A PROGRAM IN THIS STYLE

```python
#!/usr/bin/env python
import sys, re, operator, string

#
# The Quarantine class for this example
#
class TFQuarantine:
    def __init__(self, func):
        self._funcs = [func]

    def bind(self, func):
        self._funcs.append(func)
        return self

    def execute(self):
        def guard_callable(v):
            return v() if hasattr(v, '__call__') else v

        value = lambda : None
        for func in self._funcs:
            value = func(guard_callable(value))
        print(guard_callable(value))

#
# The functions
#
def get_input(arg):
    def _f():
        return sys.argv[1]
    return _f

def extract_words(path_to_file):
    def _f():
        with open(path_to_file) as f:
            data = f.read()
        pattern = re.compile('[\W_]+')
        word_list = pattern.sub(' ', data).lower().split()
        return word_list
    return _f

def remove_stop_words(word_list):
    def _f():
        with open('../stop_words.txt') as f:
            stop_words = f.read().split(',')
        # add single-letter words
        stop_words.extend(list(string.ascii_lowercase))
        return [w for w in word_list if not w in stop_words]
    return _f

def frequencies(word_list):
    word_freqs = {}
    for w in word_list:
        if w in word_freqs:
            word_freqs[w] += 1
```

```
55          else:
56              word_freqs[w] = 1
57      return word_freqs
58
59  def sort(word_freq):
60      return sorted(word_freq.items(), key=operator.itemgetter(1),
            reverse=True)
61
62  def top25_freqs(word_freqs):
63      top25 = ""
64      for tf in word_freqs[0:25]:
65          top25 += str(tf[0]) + ' - ' + str(tf[1]) + '\n'
66      return top25
67
68  #
69  # The main function
70  #
71  TFQuarantine(get_input)\
72  .bind(extract_words)\
73  .bind(remove_stop_words)\
74  .bind(frequencies)\
75  .bind(sort)\
76  .bind(top25_freqs)\
77  .execute()
```

25.3 COMMENTARY

THIS STYLE makes use of another variation of function composition. Its constraints are very interesting, namely the first one: the core program cannot do IO. For our term-frequency task, that reads files and outputs a result on the screen, this constraint poses a puzzling problem: how can we do it if the functions can't read files such as the *Pride and Prejudice* text and can't print things on the screen? Indeed, how can one write programs these days that don't interact one way or another with the user, the file system, and the network while the program is executing?

Before explaining how to do it, let's first look into *why* anyone would want to program under such seemingly unreasonable constraints. Whenever program functions need to interact with the outside world, they lose the "purity" of mathematical functions – they stop being just relations of inputs to outputs, and they either get or leak data through some other means. These "impure" functions are harder to handle, from a software engineering perspective. Take, for example, the function extract_words defined in the Passive Aggressive style (line #34 in that example). There are absolutely no guarantees that two different calls to that function with the exact same path_to_file argument produce the exact same word list: for example, someone could replace the file in between the two calls. Because of the unpredictability of the outside world, "impure" functions are more difficult than "pure" functions to reason about (to test, for example). As such, a certain philosophy of program design calls for avoiding, or at least minimizing, IO.[1] The style in this chapter is inspired by this design philosophy and Haskell's IO monad: it is a flat-out quarantine of all IO operations. Here is how it works.

The core program functions, i.e. the first-order functions, cannot do IO; they need to be "pure," in the sense that one or more calls to them with the same arguments should always produce the same results. However, higher-order functions can do IO. So the overall approach is to wrap all "IO-infected" code in higher-order functions, chain those in a contained sequence without executing them, and execute that chain only in the main program, when we absolutely need to do IO.

Let's take a look at the example program. First, let's look at the program's functions between lines #27 and #66. There are two kinds of functions: those that do IO and those that don't. The functions that do IO are: (1) get_input (lines #27–30), which reads from the command line; (2) extract_words (lines #32–39), which opens and reads a file; and (3) remove_stop_words (lines #41–48), which opens and reads the stop words file. The other three functions – frequencies, sort and top25_freqs – are "pure" in the sense that we have used that word before: given the same inputs, they will always produce the same outputs, without interaction with the outside world.

---

[1]Yes, one could write an essay with the title *IO Considered Harmful*, and someone already has, albeit in the context of CS education.

In order to identify the functions that do IO, and separate them from the rest, we have abstracted the bodies of those functions into higher-order functions:

```
1  def func(arg):
2      def _f():
3          ...body...
4      return _f
```

Doing so makes the first-order program functions be "pure," in the sense that any calls to any of them will always return the same value (their inner function) without any side effects. Here is what the Python interpreter will say if we call get_input:

```
>>> get_input(1)
<function _f at 0x01E4FC70>

>>> get_input([1, 2, 3])
<function _f at 0x01E4FC30>

>>> get_input(1)
<function _f at 0x01E4FC70>

>>> get_input([1, 2, 3])
<function _f at 0x01E4FC30>
```

We are quarantining the IO code so that it doesn't get executed at the first level of our program's functions. At this first level, these functions are again easy to deal with – they don't do anything and just return a function. Calling them is perfectly safe: nothing in the outside world will be affected, because the inner function is not being executed [yet]. The other three "pure" functions are written in the normal, direct way.

But if we are delaying the application of the IO-infected functions, how can we actually compose the functions so that they read files, count words and print characters on the screen? The first thing to notice is that this will not work:

```
1  top25_freqs(sort(frequencies(remove_stop_words(extract_words(
       get_input(None))))))
```

We need another way to define a sequence of functions. Furthermore, we need to hold on to that sequence until the time comes to interact with the outside world. As per constraint of this style, that time is in the main function: IO cannot be done in arbitrary parts of the program's execution.

The chaining is done in the main program, line #71 onward, using an instance of the Quarantine class defined in lines #7–22. We've seen these chains before in Style #9, The One. But this chain is a bit different; let's take a look at the TFQuarantine class.

Like the other monad-inspired classes that we have seen before, this class also consists of a constructor, a bind method and a third method, this time called execute, that discloses what's inside. The idea for instances of this class is to hold on to a list of functions without calling them until `execute` is called. As such, `bind` simply appends the given function to the list of functions (line #12), returning the instance for further bindings or for execution (line #13). `execute` is where the action takes place: it iterates through the list of functions (line #20), calling them one by one (line #21). The argument for each call is the return value of the previous call. At the end of the iteration, we print out the last value (line #22).

The `TFQuarantine` does *lazy evaluation* of the chain of functions: it first stores them without calling them, and only when `execute` is called in main does it call them.

In implementing `execute` we need to be careful, because, by voluntary choice, we have set ourselves for having two kinds of functions: those that return higher-order functions (the IO-infected code) and those that have normal function bodies. Since we can have both kinds of functions in these chains, the `execute` method needs to know whether it needs to apply the value or whether it simply needs to reference it (the argument in the function call in line #21). Hence the `guard_callable` inner function in lines #16–17: this function calls the value or references it, depending on whether that value is callable (functions are callable) or not (simple data types like string and dictionaries aren't callable).

It should be noted that the style in this chapter, as well as the particular implementation shown to exemplify it, don't reflect Haskell's IO monad in important ways. Faithful reproduction is not the goal here; we are focusing on the most important constraints of each style. But it's important to understand what those differences are.

First of all, Haskell is a strongly typed language, and its implementation and use of monads is very much tied to types. Functions that perform IO are of certain IO types that are part of the language. That is not the case with Python and with the implementation of the Quarantine style shown here. Haskell provides some syntactic sugar to chain functions using the do-notation, which makes these chains look like sequences of imperative commands. We didn't do that here either. More importantly, in Haskell this style is not a voluntary choice on the part of the programmers; it's part of the language design, and it's strongly enforced via type inference. That is, IO must be done this way.[2] For example, we cannot execute an IO monad in some arbitrary function; it needs to have been called from the main function. In our example, all the choices – like, for example, deciding to flag IO functions by having them return a higher-order function – are voluntary, and established with the sole purpose of making the constraints of the style visible in the code. As usual in this book, other implementation options could have been taken that would also honor the constraints.

A final comment about whether this style really achieves its ultimate purpose of minimizing IO. Clearly, it falls short. Programmers can still write

---

[2]Ignoring unsafePerformIO.

programs with as much IO as with any other style, just as we did for term frequency. However this style does one thing on the way to that ultimate goal: it forces programmers to think carefully about what functions do IO and what functions don't do IO; by thinking about it, they may be more responsible in separating IO code from the rest, and that can only be good.

25.4 THIS STYLE IN SYSTEMS DESIGN

The issue of IO being problematic goes well beyond designing programs at the small scale. In fact, its problematic nature is more visible in large distributed systems, where disk access, network latency and server load can have a tremendous effect on the user experience.

Consider, for example, a Web server that is part of a multi-user game, and that returns an image for the following kinds of URLs:

`http://example.com/images/fractal?minx=-2&maxx=1&miny=-1&maxy=1` This fractal service can be implemented in at least 2 different ways: (1) it can retrieve the image from a database where fractal images are stored, possibly generating and storing the image first if it doesn't exist on the database yet; or (2) it can always generate the image on the fly without accessing the disk. The first approach is the classical *compute & cache value* approach that we see pervasively used in computing systems, and is equivalent to "impure" functions described above. The second approach is the equivalent of "pure" functions, and seems, at first glance, worse, because for every request with the same parameters we are recomputing the image again, using more CPU cycles.

In both cases, and given that the Web is designed with explicit caches in mind, the image server can tag these images as cacheable for a very long time, which allows Web caches on the Internet to decrease the load on the original server. This weakens the argument about approach 2 being worse.

A second aspect to consider is the time of disk access vs. the time for computing the image. CPUs are so fast these days, that disk accesses have become bottlenecks in many applications. In many cases, there are substantial performance increases when using procedural generation of data vs. retrieving the pre-generated data from disk. In the case of our image server, that may or may not be the case, but if responsiveness is important, this tradeoff would need to be checked.

A third aspect to consider is the variety of images to be served and the amount of disk that it will take to store them. Our image service can generate an infinite amount of different fractal images, one for every combination of parameters `minx`, `maxx`, `miny`, and `maxy`. Images usually take up a significant number of bytes. So if we expect thousands of clients to request hundreds of thousands of different fractal images, storing them may be a bad idea.

Finally, another aspect to consider is the consequences of changes in the service's specification. For example, the first implementation of this service might generate images in the red part of the spectrum, but we may want to change it at some point to generate images in the blue part of the spectrum (say, the graphic designers changed their minds about the color scheme). If that happened using approach 1 (the database), we would need to delete these images from the database – something that may be problematic in itself, depending on how those images were stored and whether there were other images stored in the same table or not. Using approach 2, this change would be trivial to handle, since images are always generated on the fly. In either case, if we had previously set the Web cache expiration of these images to a future date, some clients would not see the changes, possibly for a long time – this shows the problem with using caches in general.[3]

If we believe that generation on the fly, i.e. "pure" functions, is beneficial for our image service, a third, even more radical, approach would be to send the server-side fractal generation function to the client, and let the client do the work, therefore unloading the server from those computations. This can only be done if that function doesn't do IO.

All this analysis goes to show that IO is, indeed, a non-trivial issue in large distributed systems. Any programming techniques and styles that shine a spotlight on this issue at the small scale are worth studying for understanding the tradeoffs at the systems design level.

25.5 HISTORICAL NOTES

Monads were brought to programming languages in the early 1990s in the context of the Haskell programming language. IO was the main reason why they were introduced, as IO has always been a contentious issue in pure functional languages.

25.6 FURTHER READING

Peyton-Jones, S. and Wadler, P. (1993). Imperative functional programming. *20th Symposium on Principles of Programming Languages* ACM Press. *Synopsis*: Another take on monads.

Wadler, P. (1997). How to declare an imperative. *ACM Computing Surveys* 29(3): 240–263. *Synopsis*: More monads. Philip Wadler's papers are always fun and interesting to read.

25.7 GLOSSARY

Pure function: A function whose result is always the same for the same input value(s), that does not depend on any data other than its explicit

---

[3] "There are only two hard things in Computer Science: cache invalidation and naming things." – quote usually attributed to Phil Karlton.

parameters and that doesn't have any observable effect in the external world.

Impure function: A function that, in addition to mapping inputs to outputs, depends on data other than its explicit parameters and/or changes the observable state of the external world.

Lazy evaluation: A program execution strategy which delays the evaluation of expressions until their values are absolutely needed.

25.8 EXERCISES

25.1 *Another language.* Implement the example program in another language, but preserve the style.

25.2 *Bound but not committed.* Find a way to demonstrate that the function chain defined in line #71 is, indeed, just creating the chain of functions without executing them.

25.3 *Top 25.* Change the `top25_freqs` function so that instead of accumulating the output on a string, it prints the data on the screen directly, one word-frequency pair at a time. Do this without violating the constraints of this style.

25.4 *True to the style.* The goal of this style is to coerce programmers into isolating their IO code from the rest. Two of the three IO-infected functions in the example program, namely `extract_words` and `remove_stop_words`, end up doing more than just IO. Refactor the program so that it does a better job at separating IO code from the rest.

25.5 *A different task.* Write one of the tasks proposed in the Prologue using this style.

VII

Data-Centric

When programming, the question *What needs to happen?* often makes us focus on functions, procedures or objects. The emphasis on algorithms in computer science reinforces that *behavior-first* approach. However, many times it's more beneficial to think of data first – that is, focus on the data of the application, and add behaviors as needed. This is a very different approach to programming, and results in different programming styles. The next three chapters show three styles that place data first and computation later. The first one, *Persistent Tables*, is the well-known relational model; the other two fall into a category of styles known as *dataflow* programming.

## CHAPTER 26

Persistent Tables

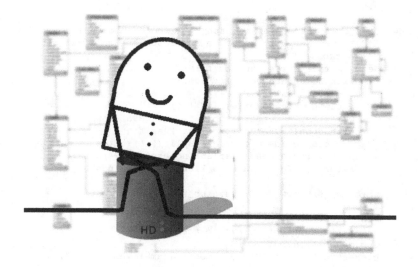

26.1 CONSTRAINTS

▷ The data exists beyond the execution of programs that use it, and is meant to be used by many different programs.

▷ The data is stored in a way that makes it easier/faster to explore. For example:

> ▷ The input data of the problem is modeled as one or more series of domains, or types, of data.

> ▷ The concrete data is modeled as having components of several domains, establishing relationships between the application's data and the domains identified.

▷ The problem is solved by issuing queries over the data.

26.2   A PROGRAM IN THIS STYLE

```python
1  #!/usr/bin/env python
2  import sys, re, string, sqlite3, os.path
3
4  #
5  # The relational database of this problem consists of 3 tables:
6  # documents, words, characters
7  #
8  def create_db_schema(connection):
9      c = connection.cursor()
10     c.execute('''CREATE TABLE documents (id INTEGER PRIMARY KEY
           AUTOINCREMENT, name)''')
11     c.execute('''CREATE TABLE words (id, doc_id, value)''')
12     c.execute('''CREATE TABLE characters (id, word_id, value)''')
13     connection.commit()
14     c.close()
15
16 def load_file_into_database(path_to_file, connection):
17     """ Takes the path to a file and loads the contents into the
           database """
18     def _extract_words(path_to_file):
19         with open(path_to_file) as f:
20             str_data = f.read()
21         pattern = re.compile('[\W_]+')
22         word_list = pattern.sub(' ', str_data).lower().split()
23         with open('../stop_words.txt') as f:
24             stop_words = f.read().split(',')
25         stop_words.extend(list(string.ascii_lowercase))
26         return [w for w in word_list if not w in stop_words]
27
28     words = _extract_words(path_to_file)
29
30     # Now let's add data to the database
31     # Add the document itself to the database
32     c = connection.cursor()
33     c.execute("INSERT INTO documents (name) VALUES (?)", (
           path_to_file,))
34     c.execute("SELECT id from documents WHERE name=?", (
           path_to_file,))
35     doc_id = c.fetchone()[0]
36
37     # Add the words to the database
38     c.execute("SELECT MAX(id) FROM words")
39     row = c.fetchone()
40     word_id = row[0]
41     if word_id == None:
42         word_id = 0
43     for w in words:
44         c.execute("INSERT INTO words VALUES (?, ?, ?)", (word_id,
               doc_id, w))
45         # Add the characters to the database
46         char_id = 0
47         for char in w:
48             c.execute("INSERT INTO characters VALUES (?, ?, ?)", (
                   char_id, word_id, char))
```

```
49          char_id += 1
50        word_id += 1
51    connection.commit()
52    c.close()
53
54 #
55 # Create if it doesn't exist
56 #
57 if not os.path.isfile('tf.db'):
58    with sqlite3.connect('tf.db') as connection:
59        create_db_schema(connection)
60        load_file_into_database(sys.argv[1], connection)
61
62 # Now, let's query
63 with sqlite3.connect('tf.db') as connection:
64    c = connection.cursor()
65    c.execute("SELECT value, COUNT(*) as C FROM words GROUP BY
                value ORDER BY C DESC")
66    for i in range(25):
67        row = c.fetchone()
68        if row != None:
69            print(row[0], '-', str(row[1]))
```

26.3 COMMENTARY

IN THIS STYLE, we want to model and store the data so that it is amenable to future retrieval in all sorts of ways. For the term-frequency task, if it's done several times, we might not always want to read and parse the file in its raw form every time we want to compute the term frequency. Also, we might want to mine for more facts about the books, not just term frequencies. So, instead of consuming the data in its raw form, this style encourages using alternative representations of the input data that make it easier to mine, now and in the future. In one approach to such a goal, the types of data (domains) that need to be stored are identified, and pieces of concrete data are associated with those domains, forming tables. With an unambiguous entity-relationship model in place, we can then fill in the tables with data, to retrieve portions of it using declarative queries.

Let's take a look at the example program, starting at the bottom. Lines #57–60 check if the database file already exists; if it doesn't exist, it creates it, and fills it out with the input file's data.

The rest of the program (from line #63 onward), which could very well be another program, queries the database. The language used to query it is the well-known Structured Query Language (SQL) used pervasively in the programming world. Line #64 gets a cursor – an object that enables traversal over records in the database (similar to an iterator in programming languages). On that cursor we then execute an SQL statement that counts the number of occurrences of each word in the words table, ordered by decreasing frequency. Finally, we iterate through the first 25 rows of the retrieved data, printing out the first column (the word) and the second one (the count). Let's look into each of the functions of the program.

create_database_schema (lines #8–14) takes the connection to the database and creates our relational schema. We divided the data into documents, words, and characters, each on a table. A document is a tuple consisting of an id (an integer) and a name; a word is a tuple consisting of an id, a cross · reference to the document id where it occurs, and a value; finally, a character is a tuple consisting of an id, a cross reference to the word id where it occurs, and a value.[1]

load_file_into_database (lines #16–52) takes a path to a file and the connection to the database and populates the tables. It first adds the document row (line #33) using the file name as the value. Lines #34–35 retrieve the automatically generated document id from the row we just inserted, so that we can use it for the words. Line #38 queries the words table for its latest word id, so that it can continue from there. Then the function proceeds to fill in the words and the characters tables (lines #43–50). Finally, the data is committed to the database (line #51) and the cursor is discarded (line #52).

---

[1]The design of entity-relationship models is a much-studied topic in CS; it is not the goal here to teach how to do it.

26.4   THIS STYLE IN SYSTEMS DESIGN

Databases are pervasive in the computing world, and relational databases, in particular, are the most popular ones. Their purpose is the same as it was back in 1955: to store data for later retrieval.

Whenever applications have this need (and they often do), there are some alternatives to the *Persistent Tables* style, but often those alternatives fall short. For starters, applications can store the data in some ad hoc manner. For simple bulk storage and retrieval, this approach may be perfectly appropriate. For example, it is quite common to see applications storing data in comma-separated value (CSV) files. However, when data is to be retrieved from storage *selectively*, rather than in bulk, better data structures need to be used. Invariably, one is led to some form of database technology, because they tend to be mature and solid pieces of software – and fast, too.

The type of database technology used depends on the application. Relational databases are good at supporting complex queries that involve many pieces of data. They take a conservative (*Tantrum*-ish) approach to adversity, by aborting any partial changes that fail to commit as a whole. Because of the ACID properties (Atomicity, Consistency, Isolation, Durability), relational databases are guaranteed to be consistent. While some applications need these features, many others don't, and more lightweight technologies, such as NoSQL, can be used instead.

In applications where there is no need to store the data for later analysis, this style of programming is, obviously, overkill.

26.5   HISTORICAL NOTES

By the early 1960s, a few companies and government laboratories were already storing and processing relatively large amounts of data, and using computers primarily as data processors. The term *database* emerged in the mid-1960s, coinciding with the availability of direct-access storage, *aka* disks – an improvement over tapes. Early on, engineers realized that storing the data in some structured manner would support faster retrieval on the new storage technology. During the 1960s, the main model used was *navigational*. A navigational database is one where the records, or objects, are found by following references from one to another. Within that model, two main approaches were being used: hierarchical databases and network databases. Hierarchical databases decompose the data into a tree, so parents can have many children but children have exactly only one parent (one-to-many); network databases extend that model to a graph.

The relational database model was formulated in the late 1960s by a computer scientist working for IBM, Edgar Codd. The ideas that came along with this model were so much better than the technology of the time, that relational databases quickly became the *de facto* standard for storing data.

In the 1980s, the emergence of object-oriented programming brought along the "object-relational impedance mismatch," the observation that the object model of OOP programs and the relational data model of long-term storage were somehow in conflict. OOP data is more like a graph, so it brought back some of the concepts of network data models of the 1960s. This mismatch gave rise to *object* and *object-relational* databases, which had some success, but not as much as one would expect. Relational databases continue to be the databases of choice these days, even when OOP languages are used.

More recently, there has been a push towards NoSQL databases, a class of data storage systems that use highly optimized key-value pairs as the basis for storage, still tabular in nature. NoSQL databases are intended for simple retrieval and appending operations, rather than retrieval of complicated data relations.

26.6 FURTHER READING

Codd, E.F. (1970). Relational model of data for large shared data banks. *Communications of the ACM* 13(6): 377–387.
Synopsis: The original paper that described the relational model and that started it all in the database field.

26.7 GLOSSARY

Entity: An N-tuple in a relational database containing data from N domains.

Relationship: An association between data and domains (a table).

26.8 EXERCISES

26.1 *Another language.* Implement the example program in another language, but preserve the style.

26.2 *Separate writer from readers.* Starting with the example program, split it into two programs: one that adds data to the database given a file, and one that reads data from the database. Change your program so that it stores the data on the file system instead of storing it in memory.

26.3 *More books.* Download another book from the Gutenberg collection, e.g. http://www.gutenberg.org/files/44534/44534-0.txt. Populate your database with both *Pride and Prejudice* and this second book.

26.4 *More queries.* Query your database to find out the answers to the following questions, and show your queries together with your answers (ignore stop words):

a. What are the 25 most frequently occurring words in each book?

b. How many words does each book have?

    c.   How many characters does each book have?

    d.   What is the longest word in each book?

    e.   What is the average number of characters per word?

    f.   What is the combined length of characters in the top 25 words of each book?

26.5 *A different task.* Write one of the tasks proposed in the Prologue using this style.

## CHAPTER 27

Spreadsheet

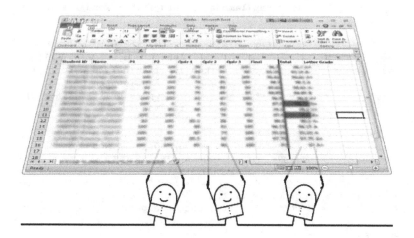

27.1 CONSTRAINTS

▷ The problem is modeled like a spreadsheet, with columns of data and formulas.

▷ Some data depends on other data according to formulas. When data changes, the dependent data also changes automatically.

27.2 A PROGRAM IN THIS STYLE

```python
#!/usr/bin/env python
import sys, re, itertools, operator

#
# The columns. Each column is a data element and a formula.
# The first 2 columns are the input data, so no formulas.
#
all_words = [(), None]
stop_words = [(), None]
non_stop_words = [(), lambda : \
                         list(map(lambda w : \
                            w if w not in stop_words[0] else '',\
                               all_words[0]))]
unique_words = [(),lambda : \
                    set([w for w in non_stop_words[0] if w!=''])]
counts = [(), lambda : \
                 list(map(lambda w, word_list : word_list.count(w),
                    \
                    unique_words[0], \
                    itertools.repeat(non_stop_words[0], \
                        len(unique_words[0])))))]
sorted_data = [(), lambda : sorted(zip(list(unique_words[0]), \
                            list(counts[0])), \
                            key=operator.itemgetter(1),
                            reverse=True)]

# The entire spreadsheet
all_columns = [all_words, stop_words, non_stop_words,\
               unique_words, counts, sorted_data]

#
# The active procedure over the columns of data.
# Call this everytime the input data changes, or periodically.
#
def update():
    global all_columns
    # Apply the formula in each column
    for c in all_columns:
        if c[1] != None:
            c[0] = c[1]()


# Load the fixed data into the first 2 columns
all_words[0] = re.findall('[a-z]{2,}', open(sys.argv[1]).read().
    lower())
stop_words[0] = set(open('../stop_words.txt').read().split(','))
# Update the columns with formulas
update()

for (w, c) in sorted_data[0][:25]:
    print(w, '-', c)
```

27.3 COMMENTARY

LIKE THE PREVIOUS STYLE, this style uses tabular data, but with a different goal in mind. Rather than storing the data and querying it later, the goal here is to emulate the good old spreadsheets that have been used in accounting for hundreds of years. In accounting, the data is laid out in tables; while some columns have primitive data, other columns have compound data that results from some combination of other columns (totals, averages, etc.). Spreadsheets are used by everyone these days; however, not many realize that their underlying programming model is quite powerful, and a great example of the dataflow family of programming styles.

Let's look at the example program. In order to understand it, it is useful to visualize an Excel (or other program) spreadsheet. The idea, conceptually, is to "place" all of the words of the book in one column, one per row, and the stop words in the second column, one per row. From then on, we have more columns that result from operations on these two columns and other columns "to the left" of each column. The complete set of columns in our spreadsheet is as follows:

1st all_words (line #8) will be filled out with all the words from the input file.

2nd stop_words (line #9) will be filled out with the words from the stop words file.

3rd non_stop_words (lines #10–13) will be filled out with all the words (from the 1st column) that aren't stop words (as given by the 2nd column).

4th unique_words (lines #14–15) will be filled out with the unique non-stop words (coming from the non_stop_words column).

5th counts (lines #16–20) will be filled out with the number of occurrences of each unique word (4th column) in the list of all non-stop words (3rd column).

6th sorted_data (lines #21–24) will be filled out with a tuple word-count sorted by count, by taking the unique words (4th column) and their counts (5th column).

Let's look closer at what these *columns* really are. Each column is modeled by a list of two elements: a list of values and a formula (a function), which may or may not exist. When the formula exists, we use the formula to generate the list of values. The update function (lines #34–39) iterates through a column at a time, setting the first element of the column structure to the result of applying the formula. This is the core update function that would need to be called periodically, or on data changes, for a long-running spreadsheet application. The example program executes the update function only once, in line #46, just after loading the input file's words into the first column (line #43) and the stop words into the second (line #44).

27.4 THIS STYLE IN SYSTEMS DESIGN

The *Spreadsheet* programming style has not been used much beyond spread-sheet applications – or at least its uses beyond spreadsheets have not been recognized as such. But the style is applicable to many more data-intensive situations.

The style is inherently declarative and reactive, meaning that it is a good fit for data-intensive applications that need a live update loop over changing data. This style is a good example of *dataflow programming*, where changes in certain points of the data space "flow" to another part of that space.

27.5 HISTORICAL NOTES

Spreadsheets were one of the first targets of computer applications, and, like so many other programming concepts, were invented by several people independently. The first spreadsheet programs were batch programs on mainframes, where the user would enter the data, press a button and wait for the rest of the data to update. The idea of making interactive spreadsheets – i.e. having the dependent data update automatically – came to life in the late 1960s with a system called LANPAR (LANguage for Programming Arrays at Random). LANPAR still used mainframes. Interactive electronic spreadsheets, this time with a GUI, were invented again in the beginning of the personal computer era in the late 1970s, with an application called VisiCalc (Visible Calculator) that ran on both the Apple computer and the PC. Spreadsheet software products have become featureful, but haven't changed much since then.

27.6 FURTHER READING

Power, D. J. (2002). A Brief History of Spreadsheets. DSSResources.com. Available at: http://dssresources.com/history/sshistory.html

27.7 GLOSSARY

Formula: A function that uses the data space in order to update a value based on other values.

27.8 EXERCISES

27.1 *Another language.* Implement the example program in another language, but preserve the style.

27.2 *Interactive.* Make the example program interactive by allowing the user to enter new file names that are then added to the data space, the columns updated, and the top 25 words displayed again.

27.3 *Column vs. cell.* The spreadsheet in the example program uses one single formula per column. Modify the program so that every cell can have its own formula.

27.4 *A different task.* Write one of the tasks proposed in the Prologue using this style.

## CHAPTER 28

Lazy Rivers

28.1 CONSTRAINTS

▷ Data is available in streams, rather than as a complete whole.

▷ Functions are filters/transformers from one kind of data stream to another.

▷ Data is processed from upstream on a need basis from downstream.

28.2   A PROGRAM IN THIS STYLE

```python
#!/usr/bin/env python
import sys, operator, string

def characters(filename):
    for line in open(filename):
        for c in line:
            yield c

def all_words(filename):
    start_char = True
    for c in characters(filename):
        if start_char == True:
            word = ""
            if c.isalnum():
                # We found the start of a word
                word = c.lower()
                start_char = False
            else: pass
        else:
            if c.isalnum():
                word += c.lower()
            else:
                # We found end of word, emit it
                start_char = True
                yield word

def non_stop_words(filename):
    stopwords = set(open('../stop_words.txt').read().split(',')  +
            list(string.ascii_lowercase))
    for w in all_words(filename):
        if not w in stopwords:
            yield w

def count_and_sort(filename):
    freqs, i = {}, 1
    for w in non_stop_words(filename):
        freqs[w] = 1 if w not in freqs else freqs[w]+1
        if i % 5000 == 0:
            yield sorted(freqs.items(), key=operator.itemgetter(1)
                , reverse=True)
        i = i+1
    yield sorted(freqs.items(), key=operator.itemgetter(1),
        reverse=True)
#
# The main function
#
for word_freqs in count_and_sort(sys.argv[1]):
    print("-----------------------------")
    for (w, c) in word_freqs[0:25]:
        print(w, '-', c)
```

28.3 COMMENTARY

THIS STYLE focuses on the problem of processing data that comes into the application continuously and may not even have an end. The same issues are seen in processing data whose size is known, but larger than the available memory. The style establishes a flow of data from upstream (data sources) to downstream (data sinks), with processing units along the way. The data flows through the stream only when the sinks need it. At any point in time, the only data present in the stream is the one needed to produce whatever piece of data the sinks need, therefore avoiding the problems raised by too much data at a time.

The example program consists of 4 functions, all of them *generators*. Generators are simplified coroutines that allow us to iterate through sequences of data on a need-basis. A generator is a function containing a `yield` statement where a `return` statement might normally be found. In the example program the data flow is established from top to bottom, textually: the top function, `characters` (lines #4–7), connects to the data source (the file) while the main instructions (lines #44–47) drive the fetching and flow of data.

Before explaining the data flow control at the bottom, let's take a look at each of the functions, from top to bottom.

- `characters` (lines #4–7) iterates through the file one line at a time (line #5). It then iterates on the line one character at a time (line #6), yielding each character downstream (line #7).

- `all_words` (lines #9–25) iterates through the characters passed to it by the function above (line #11) looking for words. The logic in this function is related to the identification of the beginning and ending of words. When the end of a word is detected, this function *yields* that word downstream (line #25).

- `non_stop_words` (lines #27–31) iterates through the words passed to it by the function above (line #29). For each word, it checks to see if it is a stop word and *yields* it only if it is not a stop word (lines #30–31).

- `count_and_sort` (lines #33–40) iterates through the non-stop words passed to it by the previous function (line #35) and increments the word counts (line #36). For every 5,000 words that it processes, it *yields* its current word-frequency dictionary, sorted (lines #37–38). The dictionary is also *yielded* at the end (line #40), because the last batch of words coming from upstream may not be an exact multiple of 5,000.

- The main instructions (lines #44–47) iterate through the word-frequency dictionaries passed by the previous function (line #44), printing them on the screen.

Functions that contain iterations yielding values are special: the next time that they are called, rather than starting from the beginning, they resume from where they yielded. So, for example, the iteration through `characters` in line #11 doesn't open the file (line #5) multiple times; instead, after the first call, every subsequent request for the next character in line #11 simply resumes the `characters` function from where it left off in line #7. The same happens in all the other generators of our program.

Having seen what each generator does, let's now look at the flow control. In line #44, there is an iteration on a sequence of word-frequency dictionaries. In each iteration, a dictionary is requested from the `count_and_sort` generator. That request prompts that generator to action: it starts iterating through the non-stop words provided to it by the `non_stop_words` generator until it gets to 5,000 words, at which point it passes the dictionary downstream. Each iteration that `count_and_sort` does through a non-stop word prompts the `non_stop_words` generator to action: it fetches the next word passed to it from upstream yielding it downstream if it is a non-stop word, or fetching another word if the one it got was a stop word. Similarly, each time that `non_stop_words` requests the next word from `all_words`, it prompts the `all_words` generator to action: it requests characters from upstream until it identifies a word, at which point it yields that word downstream.

The data flow control is, then, driven by the sink code at the bottom: data is only fetched from the source, flowing through the generators in the middle, for as long as the main instructions need it. Hence the adjective *lazy* in the name of the style, in contrast to the *eager* form of normal functions. For example, if instead of the iteration in line #44 we had this:

```
word_freqs = count_and_sort(sys.argv[1]).next()
```

only one word-frequency dictionary, the first, would be printed out, and only one portion of the input file would be read, instead of the entire file.

There are many similarities between the Lazy Rivers style and the Pipeline style described in Chapter 6. The important difference is in the way that the data flows through the functions: in the Pipeline style, the data is one single "blob" that passes from one function to another, all at once (e.g. the entire list of words); in the Lazy Rivers style, the data flows, lazily, piece by piece; it is only fetched from the source as needed by the sink.

The Lazy Rivers style is nicely expressed when programming languages support generators. Some programming languages, e.g. Java, don't support generators; the Lazy Rivers style can be implemented using *iterators* in Java, but the code will look ugly. When the programming language doesn't support generators or iterators, it is still possible to support the goals of this style, but the expression of that intent is considerably more complicated. In the absence of generators and iterators, the next best mechanism for implementing the constraints underlying this style is with threads. The style presented in the next chapter, Actors, would be a good fit for this data-centric style.

28.4 THIS STYLE IN SYSTEMS DESIGN

The *Lazy Rivers* style is of great value in data-intensive applications, especially those where the data is either live-streamed, or very large, or both. Its strength comes from the fact that only a portion of data needs to be held in memory at any point in time, that amount being driven by the end-goal need for the data.

Within components, language support for generators makes for elegant programs in this style. As will be seen in Chapter 29, threads used in a special way are a viable alternative for implementing *Lazy Rivers*. Threads, however, tend to be much heavier than generators in terms of creation and context switching.

28.5 HISTORICAL NOTES

Coroutines were first introduced in the context of a compiler for COBOL in 1963. However, they have not always been incorporated in programming languages since then. Several mainstream languages – most notably C/C++ and Java – lack support for any flavor of coroutines.

Generators were first described in the context of the language CLU circa 1977, where they were called iterators. These days, the word *iterator* is used to denote the object-oriented flavor of the concept, i.e. an object that traverses a container. The word *generator* has converged to denoting the specialized coroutines that support iteration.

28.6 FURTHER READING

Conway, M. (1963). Design of a separable transition-diagram compiler. *Communications of the ACM* 6(7): 396–408.
Synopsis: The description of the COBOL compiler design, presenting coroutines.

Liskov, B., Snyder, A., Atkinson, R. and Schaffert, C. (1977). Abstraction mechanisms in CLU. *Communications of the ACM* 20(8): 564–576.
Synopsis: This paper describes the language CLU, which featured the early concept of iterators.

28.7 GLOSSARY

Coroutine: Procedures that allow multiple entry and exit points for suspending and resuming execution.

Generator: (*aka* semicoroutine) A special kind of coroutine used to control iteration over a sequence of values. A generator always yields control back to the caller, rather than to an arbitrary place of the program.

Iterator: An object that is used to traverse a sequence of values.

28.8 EXERCISES

28.1 *Another language.* Implement the example program in another language, but preserve the style.

28.2 *Lines vs. characters.* The example program, in its eagerness to demonstrate data flows of all kinds, ends up doing something monolithic – the function `all_words` (*yuk!*). It would be much better to use Python's facilities to handle words (e.g. split). Change the program, without changing the style, so that the first generator yields an entire line, and the second yields words from those lines using the proper library functions.

28.3 *Iterators.* Some languages that don't support generators support their more verbose cousins, *iterators* (e.g. Java). Python supports both. Change the example program so that it uses iterators instead of generators.

28.4 *A different task.* Write one of the tasks proposed in the Prologue using this style.

VIII

Concurrency

The styles we have seen so far apply to all applications in general. The next four styles are specifically for applications that have concurrent units. Concurrency comes into the picture either because the applications have multiple concurrent sources of input, or because they consist of independent components distributed over a network, or because they benefit from partitioning the problem in small chunks and using the underlying multicore computers more efficiently.

## CHAPTER 29

Actors

Similar to the *Letterbox* style (Chapter 12), but where the *things* have independent threads of execution.

29.1 CONSTRAINTS

▷ The larger problem is decomposed into *things* that make sense for the problem domain.

▷ Each *thing* has a queue meant for other *things* to place messages in it.

▷ Each *thing* is a capsule of data that exposes only its ability to receive messages via the queue.

▷ Each *thing* has its own thread of execution independent of the others.

29.2 A PROGRAM IN THIS STYLE

```python
1  #!/usr/bin/env python
2
3  import sys, re, operator, string
4  from threading import Thread
5  from queue import Queue
6
7  class ActiveWFObject(Thread):
8      def __init__(self):
9          Thread.__init__(self)
10         self.name = str(type(self))
11         self.queue = Queue()
12         self._stopMe = False
13         self.start()
14
15     def run(self):
16         while not self._stopMe:
17             message = self.queue.get()
18             self._dispatch(message)
19             if message[0] == 'die':
20                 self._stopMe = True
21
22 def send(receiver, message):
23     receiver.queue.put(message)
24
25 class DataStorageManager(ActiveWFObject):
26     """ Models the contents of the file """
27     _data = ''
28
29     def _dispatch(self, message):
30         if message[0] == 'init':
31             self._init(message[1:])
32         elif message[0] == 'send_word_freqs':
33             self._process_words(message[1:])
34         else:
35             # forward
36             send(self._stop_word_manager, message)
37
38     def _init(self, message):
39         path_to_file = message[0]
40         self._stop_word_manager = message[1]
41         with open(path_to_file) as f:
42             self._data = f.read()
43         pattern = re.compile('[\W_]+')
44         self._data = pattern.sub(' ', self._data).lower()
45
46     def _process_words(self, message):
47         recipient = message[0]
48         data_str = ''.join(self._data)
49         words = data_str.split()
50         for w in words:
51             send(self._stop_word_manager, ['filter', w])
52         send(self._stop_word_manager, ['top25', recipient])
53
54 class StopWordManager(ActiveWFObject):
```

```
55      """ Models the stop word filter """
56      _stop_words = []
57
58      def _dispatch(self, message):
59          if message[0] == 'init':
60              self._init(message[1:])
61          elif message[0] == 'filter':
62              return self._filter(message[1:])
63          else:
64              # forward
65              send(self._word_freqs_manager, message)
66
67      def _init(self, message):
68          with open('../stop_words.txt') as f:
69              self._stop_words = f.read().split(',')
70          self._stop_words.extend(list(string.ascii_lowercase))
71          self._word_freqs_manager = message[0]
72
73      def _filter(self, message):
74          word = message[0]
75          if word not in self._stop_words:
76              send(self._word_freqs_manager, ['word', word])
77
78  class WordFrequencyManager(ActiveWFObject):
79      """ Keeps the word frequency data """
80      _word_freqs = {}
81
82      def _dispatch(self, message):
83          if message[0] == 'word':
84              self._increment_count(message[1:])
85          elif message[0] == 'top25':
86              self._top25(message[1:])
87
88      def _increment_count(self, message):
89          word = message[0]
90          if word in self._word_freqs:
91              self._word_freqs[word] += 1
92          else:
93              self._word_freqs[word] = 1
94
95      def _top25(self, message):
96          recipient = message[0]
97          freqs_sorted = sorted(self._word_freqs.items(), key=
                  operator.itemgetter(1), reverse=True)
98          send(recipient, ['top25', freqs_sorted])
99
100 class WordFrequencyController(ActiveWFObject):
101
102     def _dispatch(self, message):
103         if message[0] == 'run':
104             self._run(message[1:])
105         elif message[0] == 'top25':
106             self._display(message[1:])
107         else:
108             raise Exception("Message not understood " + message
                    [0])
109
```

```
110     def _run(self, message):
111         self._storage_manager = message[0]
112         send(self._storage_manager, ['send_word_freqs', self])
113
114     def _display(self, message):
115         word_freqs = message[0]
116         for (w, f) in word_freqs[0:25]:
117             print(w, '-', f)
118         send(self._storage_manager, ['die'])
119         self._stopMe = True
120
121 #
122 # The main function
123 #
124 word_freq_manager = WordFrequencyManager()
125
126 stop_word_manager = StopWordManager()
127 send(stop_word_manager, ['init', word_freq_manager])
128
129 storage_manager = DataStorageManager()
130 send(storage_manager, ['init', sys.argv[1], stop_word_manager])
131
132 wfcontroller = WordFrequencyController()
133 send(wfcontroller, ['run', storage_manager])
134
135 # Wait for the active objects to finish
136 [t.join() for t in [word_freq_manager, stop_word_manager,
        storage_manager, wfcontroller]]
```

29.3   COMMENTARY

THIS STYLE is a direct extension of the *Letterbox* style, but where the objects have their own threads. These objects are also known as *active objects* or *actors*. Objects interact with each other by sending messages that are placed in queues. Each active object performs a continuous loop over its queue, processing one message at a time, and blocking if the queue is empty.

The example program starts by defining a class, ActiveWFObject (lines #7–20), that implements generic behavior of active objects. Active objects inherit from Thread (line #7), a Python class that supports concurrent threads of execution. This means that their run method (lines #15–20) is spawned concurrently when the thread's start method is called in line #13. Each active object has a name (line #10) and a queue (line #11). The Queue object in Python implements a queue data type where threads that call the get operation may be blocked if the queue is empty. The run method (lines #15–20) runs an infinite loop that takes one message from the queue, possibly blocking if the queue is empty, and that dispatches that message. One special message die breaks the loop and makes the thread stop (lines #19–20). Any active application objects inherit the behavior from ActiveWFObject.

Lines #22–23 define a function for sending a message to a receiver. In this case, sending a message means placing it in the queue of the receiver (line #23).

Next, we have the four active application objects. In this program, we have followed the same design as that used in the *Letterbox* style (Chapter 12), so the classes and their roles are exactly the same: there is a data storage entity (lines #25–52), a stop word entity (lines #54–76), an entity for keeping word frequency counts (lines #78–98) and the application controller (lines #100–119). All of these inherit from ActiveWFObject, meaning that any instantiation of these classes spawns new threads independently running the run method (lines #15–20).

In the main method (lines #124–136), we instantiate one object of each class, so when the application runs, it results in 4 threads plus the main thread. The main thread simply blocks until the active objects' threads all stop (line #136).

A message in our program is simply a list with any number of elements that has the message tag in its first position. Object references can be sent via messages. For example, the 'init' message that the main thread sends to the StopWordManager object is ['init', word_freq_manager] (line #127), where word_freq_manager is the reference to another active object, an instance of WordFrequencyManager; the 'init' message that the main thread sends to the DataStorageManager object is ['init', sys.argv[1], stop_word_manager].

Let's look into each active application object in more detail, and the messages that are exchanged among them. The application starts by sending the 'init' message to both the stop word manager (line #127) and the data storage manager (line #130). These messages are dispatched by the corresponding active objects' threads (line #18), which results in the execution of the corresponding dispatch methods – lines #58–65 and lines #29–36, respectively. In both cases, the 'init' messages result in files being read, and its data processed in some form. Next, the main thread sends the 'run' message to the application controller (line #133), and this triggers the execution of the term frequency task over the input data. Let's see how.

Upon reception of the 'run' message, the word frequency controller stores the reference for the data storage object (line #111) and sends it the message 'send_word_freqs' with a reference to itself (line #112). In turn, when the data storage object receives 'send_word_freqs' (line #32), it starts processing the words (lines #46–52), which results in sending each word to the stop word manager object, along with the message 'filter' (lines #50–51). Upon receiving 'filter' messages, the stop word manager object filters the word (lines #73–76), which results in sending the non-stop words along with the message 'word' to the word frequency manager (lines #75–76). In turn, the word frequency manager object increments the counts for each word received via the message 'word' (lines #88–93).

When the data storage manager runs out of words, it sends the message 'top25' to the stop word manager, along with the reference to the recipient (line #52) – remember the recipient is the application controller (see line #112). The stop word manager, however, does not understand that message, as it is not one of the expected messages in its dispatch method (lines #58–65). Its dispatch method is implemented so that any message that is not explicitly expected is simply forwarded to the word frequency manager object, so the 'top25' message is forwarded. In turn, the word frequency manager understands the 'top25' message (line #85); upon its reception, it sends the sorted list of word frequencies to the recipient (lines #95–98) along with the message 'top25'. The recipient, the application controller, upon receiving the 'top25' message (line #105) prints the information on the screen (lines #115–117), and sends the 'die' message down the chain of objects, which makes them all stop (lines #18–19). At that point, all threads are finished, the main thread is unblocked, and the application ends.

Unlike the *Letterbox* style, the Actors style is inherently asynchronous, with blocking queues serving as the interfaces between the agents. The calling objects place messages in queues of the callees and continue without waiting for the dispatch of those messages.

29.4 THIS STYLE IN SYSTEMS DESIGN

This style is a natural match for large distributed systems: without distributed shared memory, components in different nodes of a network interact by sending messages to each other. There are a few ways of designing message-based systems; one of them, known as *point-to-point messaging*, where the message has a single, well-known receiver, maps directly to this style. The Java Message

Service (JMS) framework is a popular framework that supports this style, along with the publish-subscribe style described in the previous chapter. In the mobile arena, the Google Cloud Messaging for Android is another example of this style in action at planetary scale.

But this style is not just for large distributed systems. Components that consist of a single multi-threaded process also benefit from the application of this style – threaded objects with queues – as a way to limit the amount of internal concurrency.

29.5   HISTORICAL NOTES

This style targets programming for concurrent and distributed applications. The general idea is as old as the first operating systems that supported concurrency, and it emerged in several forms during the 1970s. Message-passing processes had been known to be a flexible way to structure operating systems; from the beginning, this model has co-existed with the alternative shared memory model. In the mid-1980s, Gul Agha formalized the model, giving these processes with queues the general name of *Actors*.

29.6   FURTHER READING

Agha, G. (1985). Actors: A model of concurrent computation in distributed systems. Doctoral dissertation, MIT Press.
Synopsis: This is the original work proposing the Actor model for concurrent programming.

Lauer, H. and Needham, R. (1978). On the duality of operating system structures. *Second International Symposium on Operating Systems.*
Synopsis: Long before concurrent programming was its own topic, researchers and developers were well aware of the design tradeoffs concerning communication between different units of execution. This paper presents a nice overview of message-passing vs. shared memory models.

29.7   GLOSSARY

Actor: An object with its own thread of execution, or a process node on a network. Actors have a queue to receive messages, and interact with each other only by sending messages.

Asynchronous request: A request where the requester doesn't wait for the reply, and where the reply, if any, arrives at some later point in time.

Message: A data structure carrying information from a sender to a known receiver, possibly transported via a network.

29.8 EXERCISES

29.1 *Another language.* Implement the example program in another language, but preserve the style.

29.2 *3+1 Threads.* Write another version of the example program also in the Actors style, but with only three active objects plus the main thread.

29.3 *Lazy Rivers, take 2.* Languages like Java don't have the `yield` statement explained in the Lazy Rivers style (Chapter 28). Implement the data-centric program in that chapter without using `yield`, and using the Actors style.

29.4 *A different task.* Write one of the tasks proposed in the Prologue using this style.

## CHAPTER 30

Dataspaces

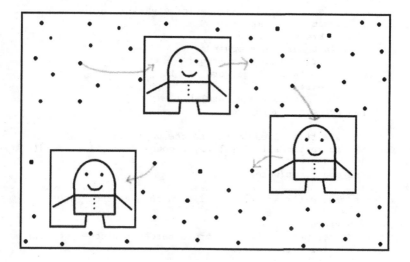

30.1 CONSTRAINTS

▷ Existence of one or more units that execute concurrently.

▷ Existence of one or more *data spaces* where concurrent units store and retrieve data.

▷ No direct data exchanges between the concurrent units, other than via the data spaces.

30.2 A PROGRAM IN THIS STYLE

```python
#!/usr/bin/env python
import re, sys, operator, queue, threading

# Two data spaces
word_space = queue.Queue()
freq_space = queue.Queue()

stopwords = set(open('../stop_words.txt').read().split(','))

# Worker function that consumes words from the word space
# and sends partial results to the frequency space
def process_words():
    word_freqs = {}
    while True:
        try:
            word = word_space.get(timeout=1)
        except queue.Empty:
            break
        if not word in stopwords:
            if word in word_freqs:
                word_freqs[word] += 1
            else:
                word_freqs[word] = 1
    freq_space.put(word_freqs)

# Let's have this thread populate the word space
for word in re.findall('[a-z]{2,}', open(sys.argv[1]).read().lower
        ()):
    word_space.put(word)

# Let's create the workers and launch them at their jobs
workers = []
for i in range(5):
    workers.append(threading.Thread(target = process_words))
[t.start() for t in workers]

# Let's wait for the workers to finish
[t.join() for t in workers]

# Let's merge the partial frequency results by consuming
# frequency data from the frequency space
word_freqs = {}
while not freq_space.empty():
    freqs = freq_space.get()
    for (k, v) in freqs.items():
        if k in word_freqs:
            count = sum(item[k] for item in [freqs, word_freqs])
        else:
            count = freqs[k]
        word_freqs[k] = count

for (w, c) in sorted(word_freqs.items(), key=operator.itemgetter
        (1), reverse=True)[:25]:
    print(w, '-', c)
```

30.3   COMMENTARY

THIS STYLE applies to concurrent and distributed systems. It is a particular kind of shared-memory style: many independently executing information processing units consume data from a common substrate and produce data onto that, or other, substrate. These substrates are called tuple, or data, spaces. There are three primitives for data manipulation: (1) **out**, or *put*, which places a piece of data from inside the unit onto a data space; (2) **in**, or *get*, which takes a piece of data from the data space and brings it into the unit; and **read**, or *sense*, which reads a piece of data from the data space into the unit without removing it.

Let's look at the example program. We use two separate data spaces, one where we place all the words (word_space, line #5), and another one where we place partial word-frequency counts (freq_space, line #6). We start by having the main thread populate the word space (lines #27–28). Then, the main thread spawns 5 worker threads and waits for them to finish (lines #31–37). Worker threads are given the process_words function (lines #12–24) to execute. This means that at this point of the program, 5 threads execute that function concurrently while the main thread waits for them to finish. So, let's look at the process_words function more closely.

The goal of the process_words function (lines #12–24) is to count word occurrences. As such, it holds on to an internal dictionary associating words with counts (line #13), and it proceeds to a loop consisting of taking one word from the word space (line #16) and incrementing the corresponding count for non-stop words (lines #19–23). The loop stops when the function is unable to get a word from the word space within 1 second (see timeout parameter in line #16), which means that there are no more words. At that point, the function simply places its internal dictionary in the frequency space (line #24).

Note that there are 5 worker threads executing process_words concurrently. This means that, very likely, different worker threads will be counting different occurrences of the same words, so each one produces only a partial word count. Given that the words are removed from the data space, no word occurrence is counted more than once.

Once the worker threads finish their jobs, the main thread unblocks (line #37) and does the rest of the computation. Its job from then on is to take the partial word counts from the frequency space and merge them into one single dictionary (lines #41–49). Finally, the information is printed on the screen (lines #51–52).

30.4   THIS STYLE IN SYSTEMS DESIGN

This style is particularly well suited for data-intensive parallelism, especially when the task scales horizontally, i.e. when the problem can be partitioned among an arbitrary number of processing units. This style can also be used in distributed systems by implementing data spaces over the network (e.g. a database). The *Dataspaces* style is not a good fit for applications where the concurrent units need to address each other.

30.5 HISTORICAL NOTES

The *Dataspaces* style was first formulated as such within the Linda programming language in the early 1980s. The model was put forward as a viable alternative to shared memory in parallel programming systems.

30.6 FURTHER READING

Ahuja, S., Carriero, N. and Gelernter, D. (1986). Linda and friends. *IEEE Computer* 19(8): 26–34.
    Synopsis: The original Linda paper that proposed the concept of tuplespaces, renamed here as dataspaces.

30.7 GLOSSARY

Tuple: Typed data object.

30.8 EXERCISES

30.1 *Another language.* Implement the example program in another language, but preserve the style.

30.2 *More concurrency.* Change the example program so that the phase of the program concerning the merging of word frequencies (lines #41–49) is done concurrently by 5 threads. Hint: think of alphabet spaces.

30.3 *A different task.* Write one of the tasks proposed in the Prologue using this style.

## CHAPTER 31

Map Reduce

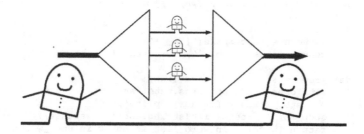

31.1 CONSTRAINTS

▷ Input data is divided in blocks.

▷ A map function applies a given worker function to each block of data, potentially in parallel.

▷ A reduce function takes the results of the many worker functions and recombines them into a coherent output.

31.2 A PROGRAM IN THIS STYLE

```python
1  #!/usr/bin/env python
2  import sys, re, operator, string
3  from functools import reduce
4  #
5  # Functions for map reduce
6  #
7  def partition(data_str, nlines):
8      """
9      Partitions the input data_str (a big string)
10     into chunks of nlines.
11     """
12     lines = data_str.split('\n')
13     for i in range(0, len(lines), nlines):
14         yield '\n'.join(lines[i:i+nlines])
15
16 def split_words(data_str):
17     """
18     Takes a string,  returns a list of pairs (word, 1),
19     one for each word in the input, so
20     [(w1, 1), (w2, 1), ..., (wn, 1)]
21     """
22     def _scan(str_data):
23         pattern = re.compile('[\W_]+')
24         return pattern.sub(' ', str_data).lower().split()
25
26     def _remove_stop_words(word_list):
27         with open('../stop_words.txt') as f:
28             stop_words = f.read().split(',')
29         stop_words.extend(list(string.ascii_lowercase))
30         return [w for w in word_list if not w in stop_words]
31
32     # The actual work of splitting the input into words
33     result = []
34     words = _remove_stop_words(_scan(data_str))
35     for w in words:
36         result.append((w, 1))
37     return result
38
39 def count_words(pairs_list_1, pairs_list_2):
40     """
41     Takes two lists of pairs of the form
42     [(w1, 1), ...]
43     and returns a list of pairs [(w1, frequency), ...],
44     where frequency is the sum of all the reported occurrences
45     """
46     mapping = {}
47     for pl in [pairs_list_1, pairs_list_2]:
48         for p in pl:
49             if p[0] in mapping:
50                 mapping[p[0]] += p[1]
51             else:
52                 mapping[p[0]] = p[1]
53     return mapping.items()
54
```

```
55 #
56 # Auxiliary functions
57 #
58 def read_file(path_to_file):
59     with open(path_to_file) as f:
60         data = f.read()
61     return data
62
63 def sort(word_freq):
64     return sorted(word_freq, key=operator.itemgetter(1), reverse=
           True)
65
66 #
67 # The main function
68 #
69 splits = map(split_words, partition(read_file(sys.argv[1]), 200))
70 word_freqs = sort(reduce(count_words, splits))
71
72 for (w, c) in word_freqs[0:25]:
73     print(w, '-', c)
```

31.3 COMMENTARY

IN THIS STYLE, the problem's input data is divided into chunks, each chunk is processed independently of the others, possibly in parallel, and the results are combined at the end. The Map Reduce style, commonly known as *MapReduce*, comprises two key abstractions: (1) a **map** function takes chunks of data, as well as a function, as arguments, and applies that function to each chunk independently, producing a collection of results; (2) a **reduce** function takes a collection of results as well as a function, as arguments, and applies that function to the collection of results in order to extract some global knowledge out of that collection.

The key observation for the term frequency task is that counting words can be done in a divide-and-conquer manner: we can count words on smaller portions of the input file (e.g. each page of the book), and then combine those counts. Not all problems can be done in this manner, but term frequency can. When this is feasible, the MapReduce solution can be very effective for very large input data, with the use of several processing units in parallel.

Let's look at the example program, starting at the bottom, lines #68–73. In line #68, the main block starts by reading the input file, partitioning it into blocks of 200 lines; those blocks are given as the second parameter to Python's map function, which takes as first parameter a worker function split_words. The result of that map is a list of partial word counts, one from each worker function, which we called splits. We then prepare those splits for reduction (line #69) – more about this later. Once ready, the splits are given as the second argument to Python's reduce function, which takes as first argument the worker function count_words (line #70). The result of that application is a list of pairs, each corresponding to a word and corresponding frequency. Let's now look into the three main functions – partition, split_words and count_words – in detail.

The partition function (lines #7–14) is a generator that takes a multi-line string and a number of lines as inputs, and generates strings with the requested number of lines. So, for example, *Pride and Prejudice* has 13,426 lines, so we are dividing it into 68 blocks of 200 lines (see line #68), with the last block having less than 200 lines. Note that the function *yields*, rather than *returns*, blocks. As seen before, this is a lazy way of processing input data, but it's functionally equivalent to returning the complete list of 68 blocks.

The split_words function (lines #16–37) takes a multi-line string – one block of 200 lines, as used in line #68 – and processes that block. The processing is similar to what we have seen before. However, this function returns its data in a very different format than that seen in other chapters for equivalent functions. After producing the list of non-stop words (lines #22–34), it iterates through that list constructing a list of pairs; the first element of each pair is a word occurrence and the second is the number 1, meaning "this word, one occurrence." For *Pride and Prejudice*'s first block, the first few entries in the resulting list look like this:

```
[('project',1),('gutenberg',1),('ebook',1),
 ('pride',1),('prejudice',1),  ('jane',1),
 ('austen',1),('ebook',1),...]
```

This seems a rather strange data structure, but it is common for MapReduce applications to make worker functions do as little computation as possible. In this case, we aren't even counting the number of occurrences of words in each block; we are simply transforming the block of words into a data structure that supports a very simple counting procedure later on.

To recapitulate, line #68 results in a list of these data, one for each of the 68 blocks.

The count_words function (lines #39–52) is the reducing worker used as the first argument to reduce in line #70. In Python, reducer functions have two arguments, which are meant to be merged in some way, and return one single value at the end. Our function takes two of the data structures described above: the first one is the result of the previous reduction, if any, starting with the empty list (line #69); the second is the new split to be merged. count_words starts by producing a dictionary out of the first list of pairs (line #46); it then iterates through the second list of pairs, incrementing the corresponding word counts in the dictionary (lines #47–51). At the end, it returns the dictionary as a list of key-value pairs. This return value is then fed into the next reduction, and the process continues until there are no more splits.

31.4   THIS STYLE IN SYSTEMS DESIGN

MapReduce is a naturally good fit for data-intensive applications where the data can be partitioned and processed independently, and the partial results recombined at the end. These applications benefit from the use of many computing units – cores, servers – that perform the mapping and reducing functions in parallel, therefore reducing the processing time by several orders of magnitude than that of a single processor. The next chapter looks into these variations of MapReduce in more detail.

Our example program, however, does not employ threads or concurrency. The example is more in line with the original LISP MapReduce. Language processors can implement the several applications of the mapped function in parallel, but that is not what Python does.[1] Nevertheless, in this book this style is grouped with the styles for concurrent programming, because those are the applications that gain the most from this style.

---

[1] Python 3.x includes a new module called concurrent.futures that provides a concurrent implementation of map.

31.5 HISTORICAL NOTES

The concept of mapping and reducing sequences, as currently used, was included in Common LISP in the late 1970s. However, those concepts predate Common LISP by at least a decade. A version of map was present in McCarthy's LISP system in 1960, under the name of maplist; this function took another function as argument that was then mapped onto each successive tail of a list argument, rather than onto each element. By the mid-1960s many dialiects of LISP had mapcar, which maps the function onto each element. Reduce was known to LISP programmers in the early 1970s. Both map and reduce were present in APL for built-in scalar operations.

Several decades later, in the early 2000s, a variation of this model was made popular by Google, who applied it at the data center scale. The model was then adopted more widely with the emergence of open source MapReduce frameworks such as Hadoop.

31.6 FURTHER READING

MAC LISP (1967). MIT A.I. Memo No.116A. Available at:
http://www.softwarepreservation.org/projects/LISP/MIT/
AIM-116A-White-Interim_User_Guide.pdf
Synopsis: This is the manual for one of the flavors of LISP, the MAC LISP, listing the functions available in that programming system. The map functions are featured prominently.

Steele, G. (1984). *Common LISP the Language*. Chapter 14.2: Concatenating, Mapping and Reducing Sequences. Digital Press. Available at:
http://www.cs.cmu.edu/Groups/AI/html/cltl/clm/clm.html
Synopsis: Common LISP had both map and reduce operations.

31.7 GLOSSARY

Map: A function takes blocks of data, as well as a function, as arguments, and applies that function to each block independently, producing a collection of results.

Reduce: A function takes a collection of results as well as a function, as arguments, and applies that function to the current merged result and the next result in the collection in order to extract some global knowledge from that collection.

31.8 EXERCISES

31.1 *Another language.* Implement the example program in another language, but preserve the style.

31.2 *Partial counts.* Change the example program so that split_words (lines #16–37) produces a list of partial word counts. Are there any advantages in doing this vs. doing what the original example program does?

31.3 *Concurrency.* Python's map and reduce functions are not multi-threaded. Write a concurrent_map function that takes a function and a list of blocks and launches a thread for each function application. Use your function instead of map in line #68. It's OK to make a few changes to the program, but try to minimize those changes.

31.4 *A different task.* Write one of the tasks proposed in the Prologue using this style.

## CHAPTER 32

Double Map Reduce

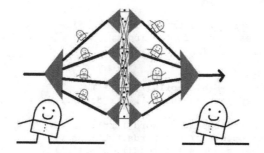

Very similar to the previous style, but with an additional twist.

32.1   CONSTRAINTS

▷ Input data is divided in blocks.

▷ A map function applies a given worker function to each block of data, potentially in parallel.

▷ The results of the many worker functions are reshuffled.

▷ The reshuffled blocks of data are given as input to a second map function that takes a reducible function as input.

▷ Optional step: a reduce function takes the results of the many worker functions and recombines them into a coherent output.

32.2 A PROGRAM IN THIS STYLE

```python
1  #!/usr/bin/env python
2  import sys, re, operator, string
3  from functools import reduce
4  #
5  # Functions for map reduce
6  #
7  def partition(data_str, nlines):
8      """
9      Partitions the input data_str (a big string)
10     into chunks of nlines.
11     """
12     lines = data_str.split('\n')
13     for i in range(0, len(lines), nlines):
14         yield '\n'.join(lines[i:i+nlines])
15
16 def split_words(data_str):
17     """
18     Takes a string, returns a list of pairs (word, 1),
19     one for each word in the input, so
20     [(w1, 1), (w2, 1), ..., (wn, 1)]
21     """
22     def _scan(str_data):
23         pattern = re.compile('[\W_]+')
24         return pattern.sub(' ', str_data).lower().split()
25
26     def _remove_stop_words(word_list):
27         with open('../stop_words.txt') as f:
28             stop_words = f.read().split(',')
29         stop_words.extend(list(string.ascii_lowercase))
30         return [w for w in word_list if not w in stop_words]
31
32     # The actual work of the mapper
33     result = []
34     words = _remove_stop_words(_scan(data_str))
35     for w in words:
36         result.append((w, 1))
37     return result
38
39 def regroup(pairs_list):
40     """
41     Takes a list of lists of pairs of the form
42     [[(w1, 1), (w2, 1), ..., (wn, 1)],
43      [(w1, 1), (w2, 1), ..., (wn, 1)],
44      ...]
45     and returns a dictionary mapping each unique word to the
46     corresponding list of pairs, so
47     { w1 : [(w1, 1), (w1, 1)...],
48       w2 : [(w2, 1), (w2, 1)...],
49       ...}
50     """
51     mapping = {}
52     for pairs in pairs_list:
53         for p in pairs:
54             if p[0] in mapping:
```

```
55                  mapping[p[0]].append(p)
56              else:
57                  mapping[p[0]] = [p]
58      return mapping
59
60  def count_words(mapping):
61      """
62      Takes a mapping of the form (word, [(word, 1), (word, 1)...)])
63      and returns a pair (word, frequency), where frequency is the
64      sum of all the reported occurrences
65      """
66      def add(x, y):
67          return x+y
68
69      return (mapping[0], reduce(add, (pair[1] for pair in mapping
            [1])))
70
71  #
72  # Auxiliary functions
73  #
74  def read_file(path_to_file):
75      with open(path_to_file) as f:
76          data = f.read()
77      return data
78
79  def sort(word_freq):
80      return sorted(word_freq, key=operator.itemgetter(1), reverse=
            True)
81
82  #
83  # The main function
84  #
85  splits = map(split_words, partition(read_file(sys.argv[1]), 200))
86  splits_per_word = regroup(splits)
87  word_freqs = sort(map(count_words, splits_per_word.items()))
88
89  for (w, c) in word_freqs[0:25]:
90      print(w, '-', c)
```

32.3 COMMENTARY

T HE BASIC MAP-REDUCE STYLE presented in the previous chapter allows for parallelization of the map step, but requires serialization of the reduce step. Hadoop, one of the most popular map-reduce frameworks, uses a slight variation that makes the reduce step also potentially parallelizable. The main idea is to regroup, or reshuffle, the list of results from the map step so that the regroupings are amenable to further mapping of a reducible function.

Let's look at the example program and how it differs from the previous one. The main function looks almost identical, but it is subtly different in two key points: (1) regrouping in line #86; and (2) the second application of map in line #87, where the previous program used reduce. Indeed, the key difference here is the regrouping of data. Let's look at it in detail.

The regroup function (lines #39–58) takes the output of the first map application as its input. Here as before, that output is a list of lists of pairs, something like this:

```
[ [('project',1),('gutenberg',1),('ebook',1),...],
  [('mr',1),('bennet',1),('among',1),...],
  ...
]
```

The purpose of the regroup function is to reorganize the data so that the counting of words can be done in parallel. It does so by reorganizing the data based on the words themselves, so that all pairs $(w_k, 1)$ end up in the same list. As an internal data structure for this, it uses a dictionary (line #51) mapping words to a list of pairs (lines #52–58). At the end, it returns that dictionary.

Once regrouped, the words are ready to be counted (lines #60–69). Counting them would be as simple as finding the size of the second element of the argument passed to count_words. The program does something slightly different: it uses a reduce function for counting (line #69). In this particular case, there is no good reason for doing it this way, and we could have done it in a more straightforward manner. However, it is important to understand that the intention of this style at this step is to *reduce* a sequence of data in some way. *Reducing* means taking a sequence of data as input and returning a smaller piece of data that merges the sequence in some way. In this case, "merging" means counting. In other cases, it could mean something else.

The important thing to note is that after regrouping (line #86), counting can be done in parallel, because we have reorganized the data per word. As such, we can apply a second map function for counting the words (line #87). With this reorganization, we could have one counting thread/processor per unique word in the file.

The style presented here is the style used by well-known MapReduce frameworks such as Hadoop, as they try to parallelize the data-intensive problems as much as possible. While certain data structures are not prone to parallel processing (e.g. the splits obtained in our program line #85), there often are transformations on that data, that make it parallelizable (e.g. the regrouping made in line #86). The process of parallelizing a complex data-intensive problem can involve several layers of regrouping.

32.4 THIS STYLE IN SYSTEMS DESIGN

At the data center scale, parallelization is done by sending blocks of data to servers that perform simple tasks. The regrouping step explained here is done by routing data to specific servers – for example, sending words starting with the letter 'a' to server s_a, 'b' to server s_b, etc.

32.5 HISTORICAL NOTES

This form of MapReduce was popularized by Google in the early 2000s. Since then, several data center-wide MapReduce frameworks have emerged, some of them open source. Hadoop is one of the most popular ones.

32.6 FURTHER READING

Dean, J. and Ghemawat, S. (2004). MapReduce: Simplified Data Processing on Large Clusters. *6th Symposium on Operating Systems Design and Implementation (ODSI'04)*.
Synopsis: Google engineers embrace MapReduce and explain how to do it at the data center scale.

32.7 EXERCISES

32.1 *Another language.* Implement the example program in another language, but preserve the style.

32.2 *You know you want to do it.* Change the example program so that count_words (lines #60–69) simply checks the length of the second element of its argument.

32.3 *Different regrouping.* Reorganizing the pairs on a per-word basis might be a bit too much parallelism! Change the program so that the regroup function reorganizes the words alphabetically into only five groups: a-e, f-j, k-o, p-t, u-z. Be mindful of what this does to the counting step.

32.4 *A different task.* Write one of the tasks proposed in the Prologue using this style.

IX

Interactivity

In all styles seen before, except the *Lazy Rivers* style in Chapter 28, the program takes input in the beginning, processes that input, and shows information on the screen at the end. Many modern applications have that characteristic, but many more have a very different nature: they take input continuously, or periodically, and update their state accordingly; there may not even be an "end of the program" as such. These applications are called *interactive*. Interaction may come from users or from other components, and it requires additional thought on how and when to update the observable output of the program. The next two chapters show two well-known styles for dealing with interactivity.

## CHAPTER 33

Trinity

33.1 CONSTRAINTS

▷ The application is divided into three parts: the model, the view, and the controller:

  ▷ the model represents the application's data;

  ▷ the view represents a specific rendition of the data;

  ▷ the controller provides for input controls, for populating/updating the model, and for invoking the right view.

▷ All application entities are associated with of one of these three parts. There should be no overlap of responsibilities.

33.2   A PROGRAM IN THIS STYLE

```python
#!/usr/bin/env python
import sys, re, operator, collections

class WordFrequenciesModel:
    """ Models the data. In this case, we're only interested
    in words and their frequencies as an end result """
    freqs = {}
    stopwords = set(open('../stop_words.txt').read().split(','))
    def __init__(self, path_to_file):
        self.update(path_to_file)

    def update(self, path_to_file):
        try:
            words = re.findall('[a-z]{2,}', open(path_to_file).
                read().lower())
            self.freqs = collections.Counter(w for w in words if w
                not in self.stopwords)
        except IOError:
            print("File not found")
            self.freqs = {}

class WordFrequenciesView:
    def __init__(self, model):
        self._model = model

    def render(self):
        sorted_freqs = sorted(self._model.freqs.items(), key=
            operator.itemgetter(1), reverse=True)
        for (w, c) in sorted_freqs[0:25]:
            print(w, '-', c)

class WordFrequencyController:
    def __init__(self, model, view):
        self._model, self._view = model, view
        view.render()

    def run(self):
        while True:
            print("Next file: ")
            sys.stdout.flush()
            filename = sys.stdin.readline().strip()
            self._model.update(filename)
            self._view.render()


m = WordFrequenciesModel(sys.argv[1])
v = WordFrequenciesView(m)
c = WordFrequencyController(m, v)
c.run()
```

33.3 COMMENTARY

T HIS STYLE is one of the most well-known styles related to interactive applications. Known as Model-View-Controller (MVC), this style embodies a general approach to architecting applications that need to report back to the user on a continuous basis. The idea is very simple and it is based on the premise that different functions/objects have different *roles*, specifically three roles. The application is divided into three parts, each containing one or more functions/objects: there is a part for modeling the data (the *model*), another part for presenting the data to the user (the *view*), and another for receiving input from the user and updating both the model and the view according to that input (the *controller*).

The main purpose of the MVC trinity is to decouple a number of application concerns, especially the model, which is usually unique, from the view and controller, of which there can be many.

The example program interacts with the user by asking her for another file after having processed the previous file. Instead of tangling algorithmic concerns with presentation concerns and user input concerns, our program does a clean separation between these three types of concerns using MVC:

- The class WordFrequenciesModel (lines #4–18) is the knowledge base of our application. Its main data structure is the term-frequency dictionary (line #7), and its main method, update, fills it out after processing the input file (lines #12–18).

- The class WordFrequenciesView (lines #20–27) is the view associated with the model. Its main method, render, gets the data from the model and prints it on the screen. We decided that presenting a sorted view of the model (line #25) is a presentation concern rather than a model concern.

- The class WordFrequenciesController (lines #29–40) runs on a loop (lines #34–40) requesting input from the user, updating the model accordingly and rendering the view to the user again.

The example program is an instance of what is known as *passive* MVC, in that the controller is the driver for both model and view updates. The passive *Trinity* style assumes that the only changes to the model are those performed by the controller. That is not always the case. Real applications often have several controllers and views, all operating on the same model, and often with concurrent actions.

Active MVC is an alternative to *passive* MVC where the view(s) are updated automatically when the model changes.[1] This can be done in a number of ways, some of them better than others. The worst way of doing it is by coupling the model with its views at construction time – e.g. sending

---

[1]Another good name for this alternative is *reactive*.

the View(s) instance(s) as an argument to the Model constructor. Reasonable implementations of *active* MVC include some version of the Hollywood style (Chapter 15) or the Actors style (Chapter 29).

The following example program uses a version of the Actors style to keep the user updated about the latest frequency counts as the file is being processed.

```python
#!/usr/bin/env python
import sys, operator, string, os, threading, re
from util import getch, cls, get_input
from time import sleep

lock = threading.Lock()


#
# The active view
#
class FreqObserver(threading.Thread):
    def __init__(self, freqs):
        threading.Thread.__init__(self)
        self.daemon, self._end = True, False
        # freqs is the part of the model to be observed
        self._freqs = freqs
        self._freqs_0 = sorted(self._freqs.items(), key=operator.
            itemgetter(1), reverse=True)[:25]
        self.start()

    def run(self):
        while not self._end:
            self._update_view()
            sleep(0.1)
        self._update_view()

    def stop(self):
        self._end = True

    def _update_view(self):
        lock.acquire()
        freqs_1 = sorted(self._freqs.items(), key=operator.
            itemgetter(1), reverse=True)[:25]
        lock.release()
        if (freqs_1 != self._freqs_0):
            self._update_display(freqs_1)
            self._freqs_0 = freqs_1

    def _update_display(self, tuples):
        def refresh_screen(data):
            # clear screen
            cls()
            print(data)
            sys.stdout.flush()

        data_str = ""
        for (w, c) in tuples:
            data_str += str(w) + ' - ' + str(c) + '\n'
        refresh_screen(data_str)
```

```
48
49  #
50  # The model
51  #
52  class WordsCounter:
53      freqs = {}
54      def count(self):
55          def non_stop_words():
56              stopwords = set(open('../stop_words.txt').read().split
                    (',') + list(string.ascii_lowercase))
57              for line in f:
58                  yield [w for w in re.findall('[a-z]{2,}', line.
                        lower()) if w not in stopwords]
59
60          words = next(non_stop_words())
61          lock.acquire()
62          for w in words:
63              self.freqs[w] = 1 if w not in self.freqs else self.
                    freqs[w]+1
64          lock.release()
65
66  #
67  # The controller
68  #
69  print("Press space bar to fetch words from the file one by one")
70  print("Press ESC to switch to automatic mode")
71  model = WordsCounter()
72  view = FreqObserver(model.freqs)
73  with open(sys.argv[1]) as f:
74      while get_input():
75          try:
76              model.count()
77          except StopIteration:
78              # Let's wait for the view thread to die gracefully
79              view.stop()
80              sleep(1)
81              break
```

A few points of this *active* MVC program are noteworthy.

First, note that the design of the program is slightly different from the first example program. For example, the controller is now just a block of code at the end (lines #69–81) rather than a class. This is inconsequential; rather than using the exact same 3 classes and method names, this second program, besides illustrating an *active* version of MVC, also makes the point that there are many different implementations honoring the same constraints. When it comes to programming styles, there are no strict laws, only constraints; it is important to be able to recognize the higher-order bits of the design of a piece of code independent of details.

Second, we have the view as an active object, with its own thread (line #11). The view object gets a freqs dictionary as input to its constructor (line #16) – this is the part of the model that it tracks. The main loop of this active object (lines #20–24) updates the internal data (line #22), showing the information to the user, and sleeps for 100 ms (line #23). Updating the internal data structures (lines #29–35) means reading the tracked word-frequencies dictionary and sorting it (line #31); it then checks whether there were changes since the last cycle (line #33) and, if so, it updates the display.

Third, our active view is not exactly like the active objects we have seen in Chapter 29; specifically, it is missing the all-important queue. The reason why the queue is missing is that in this simple program no other object is sending messages to it.

Finally, given that the view is actively polling a portion of the model every 100 ms, neither the controller nor the model need to notify the view of any changes – there is no "view, update yourself" signal anywhere. The controller still signals the model for updates (line #76).

33.4   THIS STYLE IN SYSTEMS DESIGN

A large number of frameworks for interactive applications use MVC, including Apple's iOS, countless Web frameworks, and a plethora of Graphical User Interface (GUI) libraries. It is hard not to use MVC! This style, like so many others, has the property of being so general that it serves as the backbone from which many pieces of software hang, each with their own styles, purposes and specialized roles. MVC can be applied at several scales, all the way from the application's architecture down to the design of individual classes.

The classification of code elements into model, view and controller parts is not always straightforward, and there are usually many reasonable options (as well as many unreasonable ones). Even in our trivial term-frequency example, we had options: sorting the words could be placed in the model instead of the presentation. In less trivial applications, the spectrum of options is even wider. When considering MVC for Web applications, for example, there are a number of lines by which we can divide the application's entities. We can approach the browsers as *dumb* terminals, placing the entirety of the MVC entities on the server side; or we can use rich JavaScript clients that code the view and at least part of the model on the client side; or we can have MVC both on the client and the server and establish coordination between the two; and many points in between. Whatever the division of labor between the client and the server is, it is always useful to think in terms of the model, view and controller roles, in order to tame some of the complexities of coordinating what the user sees with the backend logic and data.

33.5   HISTORICAL NOTES

MVC was first devised in 1979 in the context of Smalltalk and the emergence of GUIs.

33.6 FURTHER READING

Reenskaug, T. (1979). MODELS-VIEWS-CONTROLLERS. *The Original MVC Reports.* Available at
http://heim.ifi.uio.no/ trygver/themes/mvc/mvc-index.html
Synopsis: The original writings of MVC.

33.7 GLOSSARY

Controller: A collection of entities that receives input from the user, changing the model accordingly, and presents a view to the user.

Model: A knowledge base of the application; a collection of data and logic.

View: A visual representation of a model.

33.8 EXERCISES

33.1 *Another language.* Implement the example program in another language, but preserve the style.

33.2 *Different interactivity.* The example program interacts with the user only after having processed an entire file. Write a version of this program that interacts with the user every 5,000 non-stop words that have been processed, showing the current values of the word-frequency counts, and prompting her "More? [y/n]". If the user answers 'y', the program continues for another 5,000 words, etc.; if they answer 'n', the program asks for the next file. Make sure to separate model, view and controller code in a reasonable manner.

33.3 *Active trinity.* Transform the first example program (or the one you did for the question above) into *active* MVC using the *Hollywood* style.

33.4 *Use the queue.* In the second example program, the active view is missing the queue, because no other object sends messages to it.

- Transform the second example program into an *Actor*, with a queue. Instead of the view polling the model every 100 ms, make the model place a message in the view's queue every 100 words.

- Explain the differences and similarities between your program and the program you did in *Hollywood* style in the previous question.

- In which situations would you use the Actor version vs. the Hollywood style?

33.5 *A different task.* Write one of the tasks proposed in the Prologue using this style.

## CHAPTER 34

Restful

34.1 CONSTRAINTS

▷ Interactive: end-to-end between an active agent (e.g. a person) and a backend.

▷ Separation between client and server. Communication between the two is synchronous in the form of request–response.

▷ Statelessness communication: every request from client to server must contain all the information necessary for the server to serve the request. The server should not store context of ongoing interaction; session state is on the client.

▷ Uniform interface: clients and servers handle *resources*, which have unique identifiers. Resources are operated on with a restrictive interface consisting of creation, modification, retrieval and deletion. The result of a resource request is a hypermedia representation that also drives the application state.

34.2   A PROGRAM IN THIS STYLE

```python
1  #!/usr/bin/env python
2  import re, string, sys
3
4  with open("../stop_words.txt") as f:
5      stops = set(f.read().split(",")+list(string.ascii_lowercase))
6  # The "database"
7  data = {}
8
9  # Internal functions of the "server"-side application
10 def error_state():
11     return "Something wrong", ["get", "default", None]
12
13 # The "server"-side application handlers
14 def default_get_handler(args):
15     rep = "What would you like to do?"
16     rep += "\n1 - Quit" + "\n2 - Upload file"
17     links = {"1" : ["post", "execution", None], "2" : ["get",
               "file_form", None]}
18     return rep, links
19
20 def quit_handler(args):
21     sys.exit("Goodbye cruel world...")
22
23 def upload_get_handler(args):
24     return "Name of file to upload?", ["post", "file"]
25
26 def upload_post_handler(args):
27     def create_data(fn):
28         if fn in data:
29             return
30         word_freqs = {}
31         with open(fn) as f:
32             for w in [x.lower() for x in re.split("[^a-zA-Z]+", f.
                       read()) if len(x) > 0 and x.lower() not in stops]:
33                 word_freqs[w] = word_freqs.get(w, 0) + 1
34         wf = list(word_freqs.items())
35         data[fn] = sorted(wf,key=lambda x: x[1],reverse=True)
36
37     if args == None:
38         return error_state()
39     filename = args[0]
40     try:
41         create_data(filename)
42     except:
43         print("Unexpected error: %s" % sys.exc_info()[0])
44         return error_state()
45     return word_get_handler([filename, 0])
46
47 def word_get_handler(args):
48     def get_word(filename, word_index):
49         if word_index < len(data[filename]):
50             return data[filename][word_index]
51         else:
52             return ("no more words", 0)
```

```
53
54      filename = args[0]; word_index = args[1]
55      word_info = get_word(filename, word_index)
56      rep = '\n#{0}: {1} - {2}'.format(word_index+1, word_info[0],
            word_info[1])
57      rep += "\n\nWhat would you like to do next?"
58      rep += "\n1 - Quit" + "\n2 - Upload file"
59      rep += "\n3 - See next most-frequently occurring word"
60      links = {"1" : ["post", "execution", None],
61               "2" : ["get", "file_form", None],
62               "3" : ["get", "word", [filename, word_index+1]]}
63      return rep, links
64
65  # Handler registration
66  handlers = {"post_execution" : quit_handler,
67              "get_default" : default_get_handler,
68              "get_file_form" : upload_get_handler,
69              "post_file" : upload_post_handler,
70              "get_word" : word_get_handler }
71
72  # The "server" core
73  def handle_request(verb, uri, args):
74      def handler_key(verb, uri):
75          return verb + "_" + uri
76
77      if handler_key(verb, uri) in handlers:
78          return handlers[handler_key(verb, uri)](args)
79      else:
80          return handlers[handler_key("get", "default")](args)
81
82  # A very simple client "browser"
83  def render_and_get_input(state_representation, links):
84      print(state_representation)
85      sys.stdout.flush()
86      if type(links) is dict: # many possible next states
87          input = sys.stdin.readline().strip()
88          if input in links:
89              return links[input]
90          else:
91              return ["get", "default", None]
92      elif type(links) is list: # only one possible next state
93          if links[0] == "post": # get "form" data
94              input = sys.stdin.readline().strip()
95              links.append([input]) # add the data at the end
96              return links
97          else: # get action, don't get user input
98              return links
99      else:
100         return ["get", "default", None]
101
102 request = ["get", "default", None]
103 while True:
104     # "server"-side computation
105     state_representation, links = handle_request(*request)
106     # "client"-side computation
107     request = render_and_get_input(state_representation, links)
```

34.3 COMMENTARY

REST, REpresentational State Transfer, is an architectural style for network-based interactive applications that explains the Web. Its constraints form an interesting set of decisions whose main goals are extensibility, decentralization, interoperability, and independent component development, rather than performance.

When learning about REST, one is invariably led to the Web. Unfortunately, that approach has a few problems that hamper, rather than help, the learning process. First, it is too easy to blur the line between the architectural style (i.e. the model, a set of constraints) and the concrete Web. Second, the examples for REST that use HTTP and Web frameworks require some previous knowledge of the Web – and that's a catch–22.

REST is a *style* – a set of constraints for writing networked applications. This style is interesting in itself, independent of the fact that it captures the essence of the Web. This chapter focuses on the set of constraints stated by REST by using the same term-frequency example used throughout the book. On purpose, this chapter doesn't cover the parts of the style that pertain to the network, but it covers the main constraints of REST.

Our example program interacts with the user by presenting them options and acting on the corresponding resources. Here is an excerpt of an interaction:

```
$ python tf-33.py
   What would you like to do?
   1 - Quit
   2 - Upload file
U> 2
   Name of file to upload?
U> ../pride-and-prejudice.txt

   #1: mr - 786

   What would you like to do next?
   1 - Quit
   2 - Upload file
   3 - See next most-frequently occurring word
U> 3

   #2: elizabeth - 635

   What would you like to do next?
   1 - Quit
   2 - Upload file
   3 - See next most-frequently occurring word
```

Lines starting with U> denote user input. The words and their frequencies are presented one by one, on demand, by decreasing order of frequency. It's not hard to imagine what this interaction would look like in HTML on a browser.

Let's look at the program, starting at the bottom. Lines #102–107 are the main instructions. The program starts by creating a request (line #102). Requests in our program are lists with three elements: a method name, a resource identifier and additional data from the client (the caller) to the server (the provider) on certain operations. The request created in line #102 invokes the method GET on the default resource, and provides no additional data, because GET operations retrieve, rather than post, data on the server. The program then goes into an infinite *ping-pong* between the provider-side code and the client-side code.

In line #105, the provider is asked to handle the request.[1] As a result, the provider sends back a pair of data that we might call *hypermedia*:

- The first element of the pair is the application's state representation – i.e. some representation of the *view*, in MVC terms.

- The second element of the pair is a collection of links. These links constitute the set of possible next application states: the only possible next states are those presented to the user via these links, and it's the user who drives which state the application will go to next.

On the real Web, this pair is one unit of data in the form of HTML or XML. In our simple example, we want to avoid complicated parsing functions, so we simply split the *hypermedia* into those separate parts. This is similar to having an alternative form of HTML that would render all the information of the page without any embedded links, and show all the possible links at the bottom of the page.

In line #107, the client takes the *hypermedia* response from the server, renders it on the user's screen and returns another request, which embodies an input action from the user.

Having looked at the main interaction loop, let's now look at the provider-side code. Lines #73–80 are the request handler function. That function checks to see if there is a handler registered for the specific request and if so, it calls it; otherwise it calls the get_default handler.

The handlers of the application are registered in a dictionary just above (lines #66–70). Their keys encode the operation (GET or POST) and the resource to be operated on. As per constraint of the style, REST applications operate on *resources* using a very restrictive API consisting of retrieval (GET), creation (POST), updates (PUT) and removal (DELETE); in our case, we use only GET and POST. Also in our case, our resources consist of:

- *default*, when no resource is specified.

- *execution*, the program itself, which can be stopped per user's request.

---

[1] In Python, *a unpacks a list a into positional arguments.

- *file forms*, the data to be filled out for uploading a file.

- *files*, files.

- *words*, words.

Let's look through each one of the request handlers:

- default_handler (lines #14–18) This handler simply constructs the default *view* and default links, and returns them (line #18). The default view consists of two menu options – quit and upload a file (lines #15–16). The default links are a dictionary mapping out the possible next states of the application, of which there are two (line #17): if the user chooses option "1" (quit), the next request will be a POST on the *execution* resource with no data; if she chooses "2" (upload a file), the next request will be a GET on the *file_form* with no data. This already shines the light on what *hypermedia* is, and how it encodes the possible next states: the server is sending an encoding of where the application can go next.

- quit_handler (lines #20–21) This handler stops the program (line #21).

- upload_get_handler (lines #23–24) This handler returns a "form," which is just a textual question, and only one next possible state, a POST on the *file* resource. Note that the link, in this case, has only two parts instead of three. At this point, the server doesn't know what the user's answer is; because this is a "form," it will be up to the client to add the user's data to the request.

- upload_post_handler (lines #26–45) This handler first checks if there is, indeed, an argument given (line #38), returning an error if there isn't (line #39). When there is an argument, it is assumed to be a file name (line #40); the handler tries to create the data from the given file (lines #41–44). The function for creating the data (lines #27–36) is similar to all functions we have seen before for parsing the input file. At the end, the words are stored on the "database," which in this case is just a dictionary in memory, mapping file names to words (line #7). The handler ends by calling the word_get_handler for the given file name and word number 0 (line #45). This means that after successfully loading the file specified by the user, the upload function comes back to the user with the "page" associated with getting words in general, requesting the top word on the file that has just been uploaded.

- word_get_handler (lines #47–63) This handler gets a file name and a word index (line #54),[2] it retrieves the word from the given index of the given file from the database (line #55), and constructs the *view* listing

---

[2]On the Web, this would look like http://tf.com/word?file=...&index=...

the menu options (line #56–59), which in this case are: quit, upload a file, and get the next word. The links associated with the menu options (lines #60–62) are: POST to the *execution* for quit, GET the *file form* for another file upload, and GET the *word*. This last link is very important and it will be explained next.

In short, request handlers take eventual input data from the client, process the request, construct a *view* and a collection of links, and send these back to the client.

The link for the next word in line #62 is illustrative of one of the main constraints of REST. Consider the situation above, where the interaction presents a word, say, the 10th most frequently occurring word on the file, and is meant to be able to continue by showing the next word, word number 11. Who keeps the counter – the provider or the client? In REST, providers are meant to be unaware of the state of the interaction with the clients; the session state is to be passed to the clients. We do that by encoding the index of the next word in the link itself: [``get'', ``word'', [filename, **word_index+1**]]. Once we do that, the provider doesn't need to keep any state regarding past interactions with the client; next time, the client will simply come back with the right index.

The last part of the program is client-side code, a simple textual "browser" defined in lines #83–100. The first thing this browser does is to render the view on the screen (lines #84–85). Next, it interprets the links data structure. We have seen two kinds of link structures: a dictionary, when there are many possible next states, and a simple list, when there is only one next state (we saw this in line #24).

- When there are many possible states, the browser requests input from the user (line #87), checks whether it corresponds to one of the links (line #88) and, if so, it returns the request embodied in that link (line #89); if the link doesn't exist, it returns the default request (line #91).

- When there is only one next state, it checks whether the link is a POST (line #93), meaning that the data is a form. In this case, too, it requests input from the user (i.e. the user fills out the form, line #94), then it appends the form data to the next request (line #95) and returns the request in that link. This is the equivalent of an HTTP POST, which appends user-entered data (*aka* message body) to the request after the request header.

34.4 THIS STYLE IN SYSTEMS DESIGN

The constraints explained above embody a significant portion of the spirit of Web applications. Not all Web applications follow it; but most do.

One point of contention between the model and reality is how to handle the application state. REST calls for transferring the state to the client at every request, and using URLs that describe actions on the server without any hidden information. In many cases, that turns out to be impractical – too much information would need to be sent back and forth. In the opposite end of the spectrum, many applications hide a session identifier in cookies, a header which is not part of the resource identifier. The cookies identify the user. The server can then store the state of the user's interaction (on the database, for example), and retrieve/update that state on the server side at every request from the user. This makes the server-side application more complex and potentially less responsive, because it needs to store and manage the state of interaction with every user.

The synchronous, connectionless, request/response constraint is another interesting constraint that goes against the practice in many distributed systems. In REST, servers do not contact clients; they simply respond to clients' requests. This constraint shapes and limits the kinds of applications that are a good fit for this style. For example, real-time multi-user applications are not a good fit for REST, because they require the server to push data to clients. People have been using the Web to build these kinds of applications, using periodic client polls and long polls. Although it can be done, these applications are clearly not a good fit for the style.

The simplicity of the interface between clients and servers – *resource* identifiers and the handful of operations on them – has been one of the major strengths of the Web. This simplicity has enabled independent development of components, extensibility and interoperability that wouldn't be possible in a system with more complex interfaces.

34.5   HISTORICAL NOTES

The Web started as an open information management system for physicists to share documents. One of the main goals was for it to be extensible and decentralized. The first Web servers came online in 1991, and a few browsers were developed early on. Since then, the Web has seen an exponential growth, to become the most important software infrastructure on the Internet. The evolution of the Web has been an organic process driven by individuals and corporations, and many hits and misses. With so many commercial interests at stake, keeping the platform open and true to its original goals has sometimes been a challenge. Over the years, several corporations have tried to add features and optimizations that, although good in some respects, would violate the principles of the Web.

By the late 1990s, it was clear that, even though the Web had evolved organically and without a master plan, it had very particular architectural characteristics that made it quite different from any other large-scale networked system that had been attempted before. Many felt it was important to formalize what those characteristics were. In 2000, Roy Fielding's doctoral dissertation described the architectural style that underlies the Web, which he called REST – REpresentational State Transfer. REST is a set of constraints for writing applications that explains, to some extent, the Web. The Web diverges from REST in some points, but, for the most part, the model is quite accurate with respect to reality.

34.6 FURTHER READING

Fielding, R. (2000). Architectural Styles and the Design of Network-based Software Architectures. Doctoral dissertation, University of California, Irvine. Available at
http://www.ics.uci.edu/~fielding/pubs/dissertation/top.htm
Synopsis: Fielding's PhD dissertation explains the REST style of network applications, alternatives to it, and the constraints that it imposes.

34.7 GLOSSARY

Resource: A *thing* that can be identified.

Universal resource identifier: (URI) A unique, universally accepted, resource identifier.

Universal resource locator: (URL) A URI that encodes the location of the resource as part of its identifier.

34.8 EXERCISES

34.1 *Another language.* Implement the example program in another language, but preserve the style.

34.2 *Upwards.* The example program traverses the list of words always in the same direction: in decreasing order of frequency. Add an option in the user interaction that allows the user to see the previous word.

34.3 *A different task.* Write one of the tasks proposed in the Prologue using this style.

X

Neural Networks

Tied to supervised machine learning, the popularity of neural networks has skyrocketed in the last few years. A major factor for this was the open source release of TensorFlow in 2017, and of simplified APIs for it, such as Keras. But the concepts underlying neural networks are as old as computers. However, they were not at the center of the dominant branch of computer developments and, for some time, were even discredited. Interestingly, they embody radically different ways of thinking about computational problems, and even about computers. The next few chapters show the several new conceptual tools brought about by neural networks, with and without learning.

The examples here use Keras with the TensorFlow backend. Even though Keras has considerably elevated the abstractions for neural network programming, writing these programs feels like programming in assembly language for a very strange computer, one that is more analog than digital and that does not follow the von Neumman architecture. For that reason, and because the concepts are really different, the example programs in this part of the book cover the parts of the term frequency problem in separate but not the entire problem. This separation makes the explanations easier, as solving the complete term frequency problem in one neural network program would require coverage of an extensive number of new concepts.

By the end of these next few chapters, it will become clear that everything we've seen so far, up to this part, has been under a hidden but foundational constraint: that of *digital* computing and its *symbolic* approach to problem solving. The different styles we've seen so far are different ways of thinking about manipulating discrete symbols – characters, words, counts. Once we understand programming with neural networks, a brand new door opens up to the old and forgotten ideas of analog computing – a world not made of 0s and 1s, discrete functions, and Boolean logic, but made of real-valued numbers, continuous functions, and their calculi. The concepts floating around in this space are still closely tied to their mathematical origins, so they feel low level and... strange. The discomfort is unavoidable and necessary.

All programs in this section share the following constraints:

▷ There are only numbers. All other types of data must be converted to/from numbers.

▷ A program is a pure function, or a sequence of pure functions, that takes numbers as input and produces numbers as output. There are no side effects.

▷ Functions are neural networks, i.e. linear combinations between input and certain weights, possibly shifted by a bias, and possibly thresholded.

▷ If they are to be automatically learned from training data, the neural functions must be differentiable.

## CHAPTER 35

Dense, Shallow, under Control

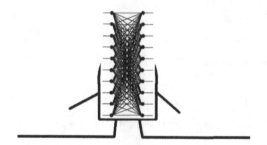

35.1 CONSTRAINTS

▷ The neural function consists of one single layer that connects all inputs to all outputs.

▷ The neural function is hardcoded by human programmers by setting the values of the weights explicitly.

35.2   A PROGRAM IN THIS STYLE

```
 1 from keras.models import Sequential
 2 from keras.layers import Dense
 3 import numpy as np
 4 import sys, os, string
 5
 6 characters = string.printable
 7 char_indices = dict((c, i) for i, c in enumerate(characters))
 8 indices_char = dict((i, c) for i, c in enumerate(characters))
 9
10 INPUT_VOCAB_SIZE = len(characters)
11
12 def encode_one_hot(line):
13     x = np.zeros((len(line), INPUT_VOCAB_SIZE))
14     for i, c in enumerate(line):
15         if c in characters:
16             index = char_indices[c]
17         else:
18             index = char_indices[' ']
19         x[i][index] = 1
20     return x
21
22 def decode_one_hot(x):
23     s = []
24     for onehot in x:
25         one_index = np.argmax(onehot)
26         s.append(indices_char[one_index])
27     return ''.join(s)
28
29 def normalization_layer_set_weights(n_layer):
30     wb = []
31     w = np.zeros((INPUT_VOCAB_SIZE, INPUT_VOCAB_SIZE), dtype=np.
          float32)
32     b = np.zeros((INPUT_VOCAB_SIZE), dtype=np.float32)
33     # Let lower case letters go through
34     for c in string.ascii_lowercase:
35         i = char_indices[c]
36         w[i, i] = 1
37     # Map capitals to lower case
38     for c in string.ascii_uppercase:
39         i = char_indices[c]
40         il = char_indices[c.lower()]
41         w[i, il] = 1
42     # Map all non-letters to space
43     sp_idx = char_indices[' ']
44     for c in [c for c in list(string.printable) if c not in list(
          string.ascii_letters)]:
45         i = char_indices[c]
46         w[i, sp_idx] = 1
47
48     wb.append(w)
49     wb.append(b)
50     n_layer.set_weights(wb)
51     return n_layer
52
```

```
53 def build_model():
54     # Normalize characters using a dense layer
55     model = Sequential()
56     dense_layer = Dense(INPUT_VOCAB_SIZE,
57                         input_shape=(INPUT_VOCAB_SIZE,),
58                         activation='softmax')
59     model.add(dense_layer)
60     return model
61
62 model = build_model()
63 model.summary()
64 normalization_layer_set_weights(model.layers[0])
65
66 with open(sys.argv[1]) as f:
67     for line in f:
68         if line.isspace(): continue
69         batch = encode_one_hot(line)
70         preds = model.predict(batch)
71         normal = decode_one_hot(preds)
72         print(normal)
```

35.3 COMMENTARY

N EURAL NETWORKS (NN) are intimately associated with supervised machine learning and, in particular, with deep learning. But these concepts are orthogonal and have emerged independently, at different times. Historically, the first learning algorithm on NNs came more than a decade after NNs were formulated. In this book, we also separate the concept of neural networks from the concept of learning from input-output examples.

This first example starts with a neural network that does not learn. Instead, it is hardcoded to behave exactly how we want it to behave. This first example is not illustrative of the kinds of programs people write with neural networks these days, but it is the simplest thing that works for explaining some of the basic concepts in *neural networks*, and setting up the stage for the concept of *supervised learning*.

The functionality here is very simple: given a sequence of characters (for example, a line), output the normalized version of those characters, where uppercase letters are converted to lowercase and every non-alphanumeric character is converted to a space. This is a simple filter designed to perform certain transformations on characters. It is also the first part of the term frequency problem.

At the center of neural networks, there is the concept of a neuron. In its mathematical model, a neuron is the realization of a function that receives N inputs, adds them together in a weighted manner, and activates a response when the resulting value meets certain conditions. The response may be a simple linear combination of the weighted inputs, but it may also be non-linear. Pictorially, the model of a neuron is as follows:

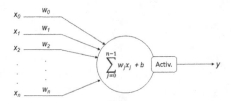

Neural networks consist of many neurons connected in some fashion. In deep learning, neurons are organized in layers, where neurons in the same layer perform the same function on the same inputs, albeit with different weights.

Before explaining the example program, it is important to note that NN programming, at least as it is currently packaged in popular frameworks, borrows many concepts from the array programming style presented in Chapter 3. If the reader skipped that chapter, now is the time to read it. The reason why NN programming is related to array programming is simple: deep learning relies heavily on the linear algebra associated with neurons and, in particular, on differentiable functions; linear algebra, in turn, relies heavily on fixed-size data – i.e. multidimensional arrays. The word *tensor* in TensorFlow refers to fixed-size multidimensional arrays that represent both data and functions over data (i.e. the layers of neurons).

A considerable part of the effort in writing NN programs, and thinking about problems in this space, falls on converting data to and from vectorized form. *Data encoding* (i.e. the vectorized representation of data) is central to NN and deep learning: certain encodings make the problem easy for the network, while others make the problem hard.

Let's take a first look at the example program, starting in its main loop in lines #66–72. That loop iterates over the lines of the given text file. For each line, it first encodes it in a special way (line #69, one-hot encoding, explained next). Then, in line #70, it puts the network to work via the model's `predict` function. Finally, it decodes the result (line #71) and prints it (line #72). The `predict` function of a network model is akin to calling the function that the network implements, for as many arguments as the number of inputs. In this case, we send it a line-sized batch of inputs, and receive a line-sized batch of outputs.

Neural networks implement linear algebra functions, and therefore can only handle scalar data. Categorical data such as characters and strings must be converted to scalar vectors before they are given as input to the network. In this case, we need to convert the characters in the text file into vectors of numbers. A popular representation for categorical data in NNs is *one-hot encoding*. This encoding is very simple: given N different things, use a vector of size N; each thing is then represented by an N-vector of N-1 zeros, and only one 1; the position of the 1 determines which thing the vector encodes.

Let's look at the one-hot encoding function in lines #12–20. This function takes a line (a string) as input and returns a 2-dimensional array of one-hot encodings, one per character in the line. The first dimension has the size of the line, so that there is one entry per character; the second dimension has the size INPUT_VOCAB_SIZE, which, in this case, is 100 (in Python, there are 100 printable characters). Each character is represented by a Numpy array of size INPUT_VOCAB_SIZE (100). All elements of that array are zero, except one, at the position corresponding to the ordinal number of the character in the set of printable characters. So, for example, the encoding for the character '0' is [1, 0, 0, ..., 0], for '1' [0, 1, 0, ..., 0], etc. For simplification, the function maps all non-printable characters to the space character (lines #17–18).

The decoding function in lines #22–27 performs the converse operation: it takes a 2-dimensional array of one-hot encoded data corresponding to characters of a line, and returns its string representation. In order to identify the character, the decoder calls Numpy's `argmax` function (line #25). Argmax returns the index of the maximum value of the given array, therefore returning the index of the single 1 in the one-hot encoded vector.

At this point, the reader might ask: why not use the ASCII or UTF-8 representation of characters, which is much shorter than 100 bits? We could do that too. The problem is that the logic of the network would be considerably more complicated to program than the logic for transforming one-hot encodings into other one-hot encodings. Let's proceed, then, to the core of the example, the neural network, also known as the *model*.

The network model is built in line #62 and printed in line #63. The model building function is defined in lines #53–60. The model is a sequence of layers (line #55), in this case only one layer. That single layer consists of a dense network (lines #56–58) that takes as input one one-hot encoded character and outputs one one-hot encoded character. A dense layer is a layer that connects every neuron in the input to every neuron in the output. The figure below shows a dense layer with 10 input neurons and 10 output neurons, for a total of 100 connections.

input

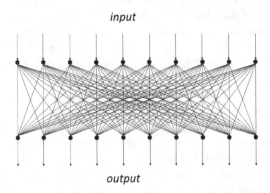

output

In the case of the example program, the dense layer connects 100 input neurons to 100 output neurons, for a total of 10,000 connections. NNs with just one layer are said to be *shallow*.

And now we come to the core of the example program: how to express the character normalization we want, using the weights of the dense layer. Normally, in NN programs, this part is *learned* from input-output examples – we will see that in the next chapter. NNs are computing machines, albeit completely different from von Neumman machines. Rather than implementing logic operations on binary data, they implement arithmetic operations on continuous signals. NNs would be better suited for analog computers. But like any computing machine, including analog computers, NNs can be explicitly programed, as long as we understand what the functions should be, and how to express it as weights on neural connections. In this case, the function – expressed as certain weight values of the dense layer – is relatively simple, and is explained next.

The "program" for the dense layer is set in line #64, and defined in lines #29–51 of the example program. The dense layer is "programmed" by setting up two parameters: the weights (line #31) and the bias (line #32), both initialized to zeros. These parameters correspond directly to the w_i's and

b depicted in the neuron figure shown a couple of pages back. The bias b is zero across the layer. The weights w is where the logic stands. First of all, it is important to understand the shape of the weights: they are a 2-dimensional matrix mapping input characters of INPUT_VOCAB_SIZE to output characters of the same size. We need to set each of these 10,000 weights so that they perform the wanted transformations. Initially, they are all zero.

Let's start with the identity function, which applies to lowercase letters (lines #34–36): for all one-hot data corresponding to lowercase letters, there should be a non-zero weight on the connection from the non-zero value of the input to the exact same position on the output. For example, the letters 'a' and 'b' correspond to the vectors $[0,0,0,0,0,0,0,0,0,1,0...,0]$ and $[0,0,0,0,0,0,0,0,0,0,1...,0]$, respectively. As such, the weights from the 10th and 11th input neurons to the 10th and 11th output neurons, respectively, are set to 1. Lines #35–36 establish this logic. With that, every time the input is 'a' or 'b', the output is also 'a' or 'b', respectively – the same happens to all lowercase letters.

The next block (lines #38–41) implements the transformation from uppercase letters to their lowercase counterparts. In this case, the weights that should not be zero are the connections from the 1-valued input neurons to the output neurons that encode the corresponding lowercase letters. For example, the letter 'A' corresponds to the vector $[35\ zeros,1,0,...,0]$. As such, there should be a non-zero weight on the connection between the 36th input neuron and the 10th output neuron, which represents the letter 'a'. Lines #39–41 establish that logic. With this, every time the input is 'A', the output is 'a' – the same happens for all other uppercase letters.

Finally, lines #43–46 map all non-letter characters to the space character. That is established by having non-zero weights on the connection between the encoding 1-value of each of those characters and the output neuron that encodes the space character.

Within the 10,000 weights of the dense layer, only 100 of them are 1; the rest are 0. Our dense layer is actually quite sparse: we could get rid of the 9,900 zero-valued connections and obtain the same behavior from the network. This calls for two observations about dense networks:

- Reevaluating the use of one-hot encoding: had we used another encoding scheme with smaller vectors, e.g. ASCII (8 bits), there would be less overall weights to set (64, specifically, for ASCII); however, the logic between input and output neurons would be a lot more complicated to express – but still feasible to implement, and left as an exercise.

- "Programming" in NNs is expressed as combinations of input values at each output neuron. Dense layers have an extensive surface area for "programming." In our case, we have 10,000 real-valued numbers at our disposal, which would allow us to capture a very large space of possible interdependencies between the input values for obtaining output values. Dense layers are the Swiss army knife of NNs – they can become anything, including more constrained networks with fewer connections, by setting certain weights to zero. We will come back to this point when analyzing networks that learn from examples.

As a final remark about the example program, the weights of the dense layer are set in line #60.

35.4 HISTORICAL NOTES

The concept of neural networks emerged within the field of theoretical neurophysiology in the 1940s. Neurophysiology studies the brain. Informed by lab experiments and empirical data, theoreticians try to capture the empirical knowledge of their field by making mathematical models that both explain the empirical results and predict behaviors not yet observed. The mathematical model of a neuron as a combinator of inputs activated by certain conditions was first described by McCulloch and Pitts in their seminal 1943 paper "A Logical Calculus of the Ideas Immanent in Nervous Activity." The paper presented neural networks, called "nervous" networks, but did not include any concept of learning. Instead, several neural networks were presented that tried to capture logic operations such as AND, OR, NOT, etc. It would be another decade until learning was introduced in NNs.

This connectionist model of computation, as it is now called, was quite different from the digital/symbolic models that influenced the development of digital computers at the time. Without their potential for learning from examples, neural networks are, indeed, much more complicated machines to program than digital circuits. The natural tendency, in this case, is that followed by McCulloch and Pitts: build logical blocks out of neurons. Doing so makes NNs no different than digital computers, and their value limited.

35.5 FURTHER READING

McCulloch, W. and Pitts, W. (1943). A logical calculus of the ideas immanent in nervous activity. In *Bulletin of Mathematical Biophysics*, Vol. 5, pp. 115–133.
Synopsis: The first paper with a mathematical model of a neuron and the concept of "nervous" networks.

35.6 GLOSSARY

Dense: A network with connections between a very large number of neurons. In Keras, a *dense layer* is a total set of connections between a set of input and a set of output neurons.

Encoding: Both a noun and a verb, it means having data in vectorized form.

Layer: Set of connections between a set of input neurons and a set of output neurons.

Model: A network architecture that is already "programmed," i.e. with all of its weights in place. The closest concept to a *program* in the NN world.

Neuron: A combinator of inputs, activated under certain conditions expressed by a continuous function.

One-hot: Popular encoding in NNs for categorical data, using one single 1 for each category, and the rest 0.

Predict: Model.Predict ≈ Program.Run ≈ Function.Eval.

Shallow: A neural network consisting just of a set of input neurons, a set of output neurons, and the connections between them; no other layers.

Tensor: Multidimensional fixed-size data representing both input/output data as well as functions (weights).

35.7 EXERCISES

35.1 *Another program.* Prove that the network "program" defined in the function `normalization_layer_set_weights` is not the only possible solution, by using different weights that produce the same character transformations.

35.2 *Another encoding.* Implement the example program using ACII encoding instead of one-hot encoding.

35.3 *Another function.* Implement an NN that transforms characters into their LEET counterparts. Use any encoding you want and any LEET code you want.

## CHAPTER 36

Dense, Shallow, out of Control

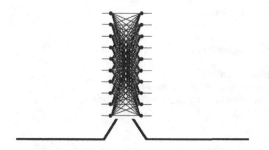

36.1   CONSTRAINTS

▷ The neural function consists of one single layer that connects all inputs to all outputs.

▷ The neural function is learned via inferences on training data.

36.2 A PROGRAM IN THIS STYLE

```python
1  from keras.models import Sequential
2  from keras.layers import Dense
3  import numpy as np
4  import sys, os, string, random
5
6  characters = string.printable
7  char_indices = dict((c, i) for i, c in enumerate(characters))
8  indices_char = dict((i, c) for i, c in enumerate(characters))
9
10 INPUT_VOCAB_SIZE = len(characters)
11 BATCH_SIZE = 200
12
13 def encode_one_hot(line):
14     x = np.zeros((len(line), INPUT_VOCAB_SIZE))
15     for i, c in enumerate(line):
16         if c in characters:
17             index = char_indices[c]
18         else:
19             index = char_indices[' ']
20         x[i][index] = 1
21     return x
22
23 def decode_one_hot(x):
24     s = []
25     for onehot in x:
26         one_index = np.argmax(onehot)
27         s.append(indices_char[one_index])
28     return ''.join(s)
29
30 def build_model():
31     # Normalize characters using a dense layer
32     model = Sequential()
33     dense_layer = Dense(INPUT_VOCAB_SIZE,
34                         input_shape=(INPUT_VOCAB_SIZE,),
35                         activation='softmax')
36     model.add(dense_layer)
37     return model
38
39 def input_generator(nsamples):
40     def generate_line():
41         inline = []; outline = []
42         for _ in range(nsamples):
43             c = random.choice(characters)
44             expected = c.lower() if c in string.ascii_letters else
                          ' '
45             inline.append(c); outline.append(expected)
46         return ''.join(inline), ''.join(outline)
47
48     while True:
49         input_data, expected = generate_line()
50         data_in = encode_one_hot(input_data)
51         data_out = encode_one_hot(expected)
52         yield data_in, data_out
53
```

```
54 def train(model):
55     model.compile(loss='categorical_crossentropy',
56                   optimizer='adam',
57                   metrics=['accuracy'])
58     input_gen = input_generator(BATCH_SIZE)
59     validation_gen = input_generator(BATCH_SIZE)
60     model.fit_generator(input_gen,
61             epochs = 50, workers=1,
62             steps_per_epoch = 20,
63             validation_data = validation_gen,
64             validation_steps = 10)
65
66 model = build_model()
67 model.summary()
68 train(model)
69
70 input("Network has been trained. Press <Enter> to run program.")
71 with open(sys.argv[1]) as f:
72     for line in f:
73         if line.isspace(): continue
74         batch = encode_one_hot(line)
75         preds = model.predict(batch)
76         normal = decode_one_hot(preds)
77         print(normal)
```

36.3 COMMENTARY

HAVING programmed a neural network explicitly in the previous chapter by manually setting the weights of the dense layer, we now turn our attention to *learning*. Learning relates to the following question: can the weights be, somehow, automatically calculated based on input-output examples? It turns out that they can. This was the important discovery that made neural networks a topic of interest to computer scientists over the years. And while the first *learning algorithms*, as they are called, weren't that good, and almost brought the work on neural networks to a halt, developments in the early 1980s changed the field.

The example program in this chapter is similar to the program of the previous chapter. The only difference is the process by which the dense layer is "programmed." In the previous chapter, the weights were part of the logic of the program; in here, the weights are learned. Let's dive in.

First, the parts that are exactly the same as the previous program are: the one-hot encode and decode functions (lines #13–28), the model definition function (lines #39–46), and the main block (lines #71–77).

What is different in this program is that, instead of setting the weights, there is now a step of *training* the network after building it (line #68, and function in lines #54–64). This is where the weights of the dense layer are calculated. As a high-level API, Keras hides most of the details of how learning works, but programmers still need to have some basic domain knowledge about machine learning in order to use Keras effectively. In here, we explain how learning *could* work in the simple network of this simple example. Note that this is not how learning actually works in TensorFlow and all other modern deep learning frameworks, but it's very close to how it was first proposed in 1958, with additional simplifications coming from the nature of the problem at hand, and the one-hot encoding that is being used. Here is a simple learning algorithm that could work in this case:

1. Initialize the weights to 0.

2. Get one one-hot encoded character from the training set. Feed this input to the network and get the output.

3. For each output neuron, if its value is 0 when it should have been 1, change the weight to 1 for the [single] input that was 1.

4. Repeat steps 2–4 until there are no more mistakes.

After seeing all possible characters as input, the network will correctly have "learned" one possible solution for the problem, which happens to be the exact same solution of the previous chapter.

This learning algorithm is too simple to work beyond the simple translation and encoding of our problem; it will not even work for other encodings of characters. Modern machine learning uses an algorithm called *backpropagation*, which is a generalization of the basic idea of propagating the error backwards in the network over many layers, not just one. Whether to increase or decrease the values of the weights, and by how much, is established by optimization algorithms, typically variations of *gradient descent*. The adjustment of the weights is done iteratively, with small changes over the training period, rather than the big change from 0 to 1 of the simplistic algorithm outlined above.

Let's get back to the example program. The training function starts by "compiling" the network (lines #55–57) for the given loss function (categorical entropy), optimizer (adam), and success metrics (accuracy). "Compiling" here means something very different than for the rest of the programming world: it means setting up and configuring the network for training. During training, the tensor backend needs to know how to optimize the parameter values given the data, and how to measure success. Loss functions (or objective functions) map the losses into a scalar value, so that the loss can be calculated and assessed. Examples of loss functions include mean squared error, binary cross-entropy, and many others. Categorical cross-entropy, used here, is appropriate for when there are two or more label classes on the output, and the classes are encoded using one-hot representation. That is the case here – each character is a category, and is encoded using one-hot representation. As for the optimizer, there are many variations for implementing gradient descent on batches of data; the one used here (adam) converges very quickly on this data. Finally, the last parameter, metrics, defines a set of metrics that should be used to measure success of the learning process. In this case, we are interested in accuracy, i.e. the distance between the true values and the predicted values during training.

Having compiled the model, the training function proceeds to the actual training, which is done in lines #60–64 by calling the fit_generator method on the model. fit_generator is a variant of the simple fit method, which fits the model to the given training data – that's how learning happens; fit_generator is fit but where the training and validation data are fed using generator functions, rather than being loaded in memory as a whole. The parameters to the fit method also require some knowledge of machine learning. Basically, learning happens on batches of the training data, and with multiple passes (called *epochs*) over it. Our batch size is 200 samples (line #11), and we're training on 50 epochs. We define steps_per_epoch to be 20, which means that the training set size is 20 * 200 = 4,000 samples. The following picture illustrates the relations among all these training concepts:

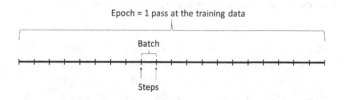

Finally, we come to the training data. For this particular problem, we can generate infinite amounts of training data, because we know exactly how to implement the specified character normalization using a traditional program. In real applications of machine learning, however, that is not the case, and training data is typically hard to come by. In this example program, the generation of training data is done in lines #39–52. In general, training data consists of pairs of input-output values, properly encoded for the model. In this case, it's batches of characters and their normalized counterparts, using one-hot representations.

There is no strong reason for using a generator vs. a normal function that would store the training data in memory. Typically, generators are preferred when the training data is too big, and does not fit in memory, but that is not the case here. The preference for the generator is that it makes it somewhat easier to experiment with different training parameters of the `fit` method (line #60) without having to change the data generation part.

At this point, the reader should have one burning question: did this network learn to do character normalization in exactly the same way we did it in the previous chapter? When we had control of the program, we set certain weights to 1, and left all others at 0. Was that what happened here too? Not exactly, but the learned solutions have similarities to ours. We can inspect the weights of any connection in any network layer with this statement:

```
print(model.layers[n].get_weights()[0][i][j])
```

where n is the layer's ordinal number, and i and j are the ordinal numbers of an input and output neuron, respectively (`get_weights()` returns a list of two items: the weights and the bias; hence the index [0]). Inspecting, for example, the weights coming out of input neuron 36 ('A') yields the following values, in the case of the programmed network of the previous chapter:

```
[0. 0. 0. 0. 0. 0. 0. 0. 0. 0. 1. 0. 0. 0. 0. 0. 0. 0. 0.
 0. 0. 0. 0. 0. 0. 0. 0. 0. 0. 0. 0. 0. 0. 0. 0. 0. 0. 0.
 0. 0. 0. 0. 0. 0. 0. 0. 0. 0. 0. 0. 0. 0. 0. 0. 0. 0. 0.
 0. 0. 0. 0. 0. 0. 0. 0. 0. 0. 0. 0. 0. 0. 0. 0. 0. 0. 0.
 0. 0. 0. 0. 0. 0. 0. 0. 0. 0. 0. 0. 0. 0. 0. 0. 0. 0. 0.
 0. 0. 0. 0. 0.]
```

The single 1 at position 10 is the programmed mapping to lowercase 'a', which is character number 10. In the case of the learned network of this chapter, the values are something like this (the exact values vary from run to run):

```
[-0.72 -0.60 -0.81 -0.63 -0.91 -0.79 -0.80 -0.60 -0.74
 -0.75  0.92 -1.03 -0.92 -1.07 -0.81 -0.92 -0.81 -0.90
 -0.79 -0.90 -1.04 -0.82 -0.99 -0.90 -1.09 -1.07 -1.00
 -0.91 -0.92 -0.90 -1.07 -0.84 -0.87 -1.08 -0.85 -1.09
```

```
-0.68 -0.65 -0.61 -0.66 -0.90 -0.63 -0.87 -0.61 -0.85
-0.72 -0.79 -0.87 -0.78 -0.57 -0.78 -0.66 -0.64 -0.72
-0.57 -0.60 -0.83 -0.59 -0.89 -0.64 -0.73 -0.82 -0.87
-0.61 -0.85 -0.82 -0.66 -0.85 -0.58 -0.60 -0.69 -0.72
-0.65 -0.79 -0.75 -0.83 -0.72 -0.62 -0.82 -0.80 -0.69
-0.62 -0.81 -0.68 -0.66 -0.58 -0.57 -0.86 -0.61 -0.80
-0.63 -0.72 -0.81 -0.74 -0.88 -0.57 -0.66 -0.75 -0.58
-0.72]
```

All values are negative, except the value at position 10, which is strongly positive. The same is true for all other cases of our programmed mappings. So even though the network does not learn the exact solution we had, it is able to learn equally valid solutions. As mentioned before, a dense layer like this one, with 10,000 real-valued numbers, has an extensive surface area for "programming." There is a large number of possible solutions, and these numbers also work for the character transformations at hand.

There is no question that the possibility of automatically learning programs from analyzing input-output examples is an exciting new capability brought by neural networks that makes them more powerful than traditional computers. The constraint that enables this new capability is the use of *differentiable functions* – functions for which it is possible to calculate their derivatives. The cost, however, at least for the time being, is the intelligibility of the resulting programs. Expressed as multidimensional arrays of real-valued numbers, programs in this style, especially when learned from data, are extremely hard to interpret, in general.

36.4   HISTORICAL NOTES

Developments in neuropsychology were relatively slow after McCulloch and Pitts' 1943 paper on "nervous" networks. In 1949, Donald Hebb published a highly influential book presenting a theory for how the brain might process and store information. This theory included the first vague ideas about learning via adjustments at the synapses (i.e. the links between neurons).

It wasn't until 1958 that the first learning algorithm was devised, and only for one single neuron: the perceptron, by Frank Rosenblatt. A perceptron is a neuron capable of learning from examples. Rosenblatt's work built on both McCulloch and Pitts' neural model of logical functions and on Hebb's vague ideas of adjustable synapses. Mathematically, he modeled the synapses as weights on the inputs of the neuron, and he came up with the idea of a training dataset. His learning algorithm was very simple:

1. Start with random weights.

2. Take one input from the dataset, feed it to the perceptron, and compute the output.

3. If the output does not match the expected result: (a) if it is 1 but it should have been 0, decrease the weights that had an input of 1; (b) if it is a 0 but should have been a 1, increase the weights that had an input of 1.

4. Get another input from the dataset, and repeat steps 2–4, until the perceptron shows no mistakes.

Rosenblatt not only devised this, but also built his design in custom hardware, showing that it could learn to classify simple shapes correctly on 20x20 pixel-like images. This achievement is seen as the birth of machine learning as a computational field.

It is believed that Marvin Minsky, considered the father of Artificial Intelligence, had done prior work in 1951 on a neural machine called SNARC (Stochastic Neural Analog Reinforcement Calculator). But there is no trace of this work, other than word-of-mouth. Interestingly, Marvin Minsky was highly skeptical of Rosenblatt's approach to machine intelligence. Part of his concerns were valid. The perceptron works well when there is only one neuron and a finite set of output values, such as 0 and 1, which is a simple classification problem. It is also possible to extend this basic algorithm to a set of perceptrons, i.e. a layer, for solving slightly more complex problems, such as the one presented in this chapter. But the perceptron has many limitations. In particular, Minsky and Papert showed that it is not possible to implement the exclusive-or (XOR) logical function with just one layer of perceptrons – that there needs to be more than one layer of perceptrons. Rosenblatt's algorithm did not work for multiple layers. Not being able to do XOR, the prospects of perceptrons as the basis for machine intelligence were deemed slim.

Partly due to Minsky's influence, and his negative views on perceptrons, the work on neural networks was virtually abandoned as a discredited dead-end – until the 1980s, when the hype around rule-based AI started to deflate, and there was space again to talk about neural networks.

36.5 FURTHER READING

Hebb, D.O. (1949). *The Organization of Behavior: A Neuropsychological Theory*. John Wiley & Sons.
Synopsis: The first vague ideas about learning in neural networks via synaptic adjustments.

Rosenblatt, F. (1958). The perceptron: a probabilistic model for information storage and organization in the brain. In *Psychological Review*, Vol. 65(6):386–408.
Synopsis: The first learning algorithm for a single neuron.

36.6   GLOSSARY

Batch: A subset of the training data used to update the learned weights.

Epoch: One pass at the entire training data.

Fit: Learn the set of weights that best fit the given input-output data.

Learning algorithm: Any algorithm for finding the weights of a neural network that minimize the error between true values and predicted values.

Loss function: Function that maps a series of errors into one single number.

Optimizer: Concrete implementations of gradient descent.

Training: A phase in neural network programming where the weights of the network are learned from the data.

Validation: A separate phase for testing how well the trained model performs on unseen data.

36.7   EXERCISES

36.1 *Learning algorithm.* Implement the simple learning algorithm outlined in the Commentary section of this chapter (page 290), and show that it works correctly.

36.2 *Other training parameters.* Read through the Keras documentation for the `compile` method, and experiment with different optimizers and loss functions in the example program. Experiment also with different number of epochs and steps per epoch. Turn in a report with your findings.

36.3 *Another encoding.* Implement the example program using ASCII encoding instead of one-hot encoding.

36.4 *Another function.* Implement an NN that learns to transform characters into their LEET counterparts. Use any encoding you want and any LEET code you want.

Bow Tie

37.1 CONSTRAINTS

▷ The shape of the network resembles a bow tie with, at least, one hidden layer.

37.2   A PROGRAM IN THIS STYLE

```
1  from keras.models import Sequential
2  from keras.layers import Dense
3  import numpy as np
4  import sys, os, string
5
6  characters = string.printable
7  char_indices = dict((c, i) for i, c in enumerate(characters))
8  indices_char = dict((i, c) for i, c in enumerate(characters))
9
10 INPUT_VOCAB_SIZE = len(characters)
11
12 def encode_one_hot(line):
13     x = np.zeros((len(line), INPUT_VOCAB_SIZE))
14     for i, c in enumerate(line):
15         index = char_indices[c] if c in characters else
                char_indices[' ']
16         x[i][index] = 1
17     return x
18
19 def decode_values(x):
20     s = []
21     for onehot in x:
22         # Find the index of the value closest to 1
23         one_index = (np.abs(onehot - 1.0)).argmin()
24         s.append(indices_char[one_index])
25     return ''.join(s)
26
27 def layer0_set_weights(n_layer):
28     wb = []
29     w = np.zeros((INPUT_VOCAB_SIZE, 1), dtype=np.float32)
30     b = np.zeros((1), dtype=np.float32)
31     # Let lower case letters go through
32     for c in string.ascii_lowercase:
33         i = char_indices[c]
34         w[i, 0] = 1.0/i
35     # Map capitals to lower case
36     for c in string.ascii_uppercase:
37         i = char_indices[c]
38         il = char_indices[c.lower()]
39         w[i, 0] = 1.0/il
40     # Map all non-letters to space
41     sp_idx = char_indices[' ']
42     for c in [c for c in list(string.printable) if c not in list(
            string.ascii_letters)]:
43         i = char_indices[c]
44         w[i, 0] = 1.0/sp_idx
45
46     wb.append(w)
47     wb.append(b)
48     n_layer.set_weights(wb)
49     return n_layer
50
51 def layer1_set_weights(n_layer):
52     wb = []
```

```
53      w = np.zeros((1, INPUT_VOCAB_SIZE), dtype=np.float32)
54      b = np.zeros((INPUT_VOCAB_SIZE), dtype=np.float32)
55      # Recover the lower case letters
56      for c in string.ascii_lowercase:
57          i = char_indices[c]
58          w[0, i] = i
59      # Recover the space
60      sp_idx = char_indices[' ']
61      w[0, sp_idx] = sp_idx
62
63      wb.append(w)
64      wb.append(b)
65      n_layer.set_weights(wb)
66      return n_layer
67
68  def build_model():
69      model = Sequential()
70      model.add(Dense(1, input_shape=(INPUT_VOCAB_SIZE,)))
71      model.add(Dense(INPUT_VOCAB_SIZE))
72      return model
73
74  model = build_model()
75  model.summary()
76  layer0_set_weights(model.layers[0])
77  layer1_set_weights(model.layers[1])
78
79  with open(sys.argv[1]) as f:
80      for line in f:
81          if line.isspace(): continue
82          batch = encode_one_hot(line)
83          preds = model.predict(batch)
84          normal = decode_values(preds)
85          print(normal)
```

37.3 COMMENTARY

THE TWO neural network programs in the previous chapters, with and without learning, stay close to discrete symbol manipulation: the characters are converted to/from arrays of numbers, but those numbers are still 0s and 1s. In this chapter, we are going to perform the same character transformations by taking advantage of the fact that we have the whole spectrum of real numbers at our disposal.

Rather than having one fully connected dense layer, we are going to have a network with a bow tie shape, like so:

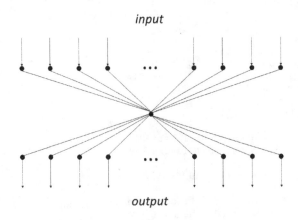

input

output

In general, bow tie networks consist of several dense *hidden layers*, where the first few layers map the multidimensional input into increasingly smaller dimensions, and the last few layers do the opposite. These networks are said to follow the *encoder-decoder* architecture; we'll see why. The bow tie network in this example program has exactly one hidden layer with exactly one neuron, like in the figure above. The problem, then is one of *encoding* the input of 100 dimensions of 0s and 1s into one real number, and then *decoding* that real number back to 100 dimensions that can be interpreted as characters. First, we are going to do it by manually setting up the weights of the two layers. Then, we will show how the weights can be learned.

Let's take a look at the example program, starting at the bottom. The model is defined in lines #68–72. It is exactly the network topology of the figure above: 100 (INPUT_VOCAB_SIZE) neurons as input, 1 neuron in the hidden layer – that's the 1 in line #70, – and 100 neurons as the output. Note that this network is very small, consisting of one pair of 100 connections, so 200 weights. Compare this to the number of weights in the previous chapters (10,000).

The model is built and its weights are set in lines #74–77. How do we set the weights so that they transform the characters in the right way? An even more basic question would be: how do we set the weights so that we get the same characters in the output, without any transformation? The identity function over a bow tie network is known as *auto-encoding*, and it has all sorts of interesting applications. Understanding the encoding from one-hot representation into a real number, and the decoding from that real number back to something that can be interpreted as a character is the key to understanding these bow tie networks.

The code for that is in lines #27–49 (weights of the first layer) and #51–66 (weights of the second layer). There is a large number of values that can be used, and that work equally well. The weights of the first and second layer just need to obey a few constraints. The basic idea used in this program is the following: given a one-hot input for character number K, where K is a positive integer between 0 and 99 (in our case), transform it into a real number by using some function, for example, the inverse $1/K$. That is the weight of each connection between the input and the hidden neuron. So, for example, 'a' and 'b' are numbers 10 and 11, respectively; the weights from the 1-encoding neurons for 'a' and 'b' to the middle layer will be 0.1 and 0.0909, respectively. That means that when the input is an 'a', the value of the hidden neuron will be 0.1; when the input is a 'b', it will be 0.0909; etc. This part is straightforward.

But since we want to perform some character transformations that aren't the identity function, we switch the weights for characters that are not lowercase letters. In lines #36–39, the connections between all uppercase neurons and the hidden neuron get assigned the weight of the lowercase counterpart. And in lines #41–44, all other characters that are not letters get assigned the weight of the space character. Again, this part is simple.

The challenge is to decode the value of the real number back into characters. One way of doing that is to set the weights of the second layer as the inverse of the first. So, for example, the weight of the connection from the hidden layer to the 'a' encoding output neuron would be 10; the weight to the 'b' encoding output neuron would be 11; etc. That way, an 'a' input produces 0.1 in the hidden neuron, and produces a 1 again in the 1-encoding neuron for 'a'. However, there is a problem: that same hidden value, 0.1, will flow through all connections of the hidden neuron to the output neurons, and none of those weights is zero. That means that the output of an 'a' will be exactly 1 in the 'a'-encoding neuron, and ninety-nine non-zero values elsewhere. For example, when the input is 'a', the 'b'-encoding output neuron will be 0.1× 11 = 1.1. In other words, the output is no longer a one-hot representation of characters.

In general, it is not possible to preserve one-hot encodings on the output by doing the kind of compression we did, using only one decoding layer to decompress it. The compression here destroys the independence of the input dimensions, and it is not possible to recover that independence with just one decoding layer. This is another manifestation of the XOR problem of perceptrons mentioned in the previous chapter. We will come back to this later on in the chapter, when we look at the learned version of this program. For now, let's accept the fact that the output is no longer one-hot.

In spite of the output not being one-hot, it is perfectly possible to recover the information about which character it encodes: we should look for the neuron whose value is closest to, or exactly, 1. That neuron is guaranteed to encode the character that went through the right pair of encoding-decoding transformations. All other non-1 values are side effects.

The example program has a new function for decoding the output: decode_values, in lines #19–25 (instead of decode_one_hot). That function looks for the index of the value that is closest to 1 – that is the index of the character we want.

Let's now turn our attention to learning. Can these weights be learned? The answer is yes, they can. In fact, we can learn different things here. Let's start with learning the exact function implemented by the example program, the one that takes one-hot representations as input and produces a vector of size 100, with real numbers, where only one value is exactly 1. The following snippet shows the most important parts of the learning counterpart to our example program:

```
1  # ...initial block...
2
3  BATCH_SIZE = 200
4
5  def encode_values(line):
6      x = np.zeros((len(line), INPUT_VOCAB_SIZE))
7      for i, c in enumerate(line):
8          index = char_indices[c] if c in characters else
                  char_indices[' ']
9          for a_c in characters:
10             if a_c == c:
11                 x[i][index] = 1
12             else:
13                 idx = char_indices[a_c]
14                 x[i][idx] = idx/index
15     return x
16
17 def input_generator(nsamples):
18     def generate_line():
19         inline = []; outline = []
20         for _ in range(nsamples):
21             c = random.choice(characters)
22             expected = c.lower() if c in string.ascii_letters else
                      ' '
23             inline.append(c); outline.append(expected)
24         return ''.join(inline), ''.join(outline)
25
26     while True:
27         input_data, expected = generate_line()
28         data_in = encode_one_hot(input_data)
29         data_out = encode_values(expected)
30         yield data_in, data_out
31
32 def train(model):
33     model.compile(loss='mse',
34                   optimizer='adam',
```

```
35              metrics=['accuracy', 'mse'])
36   input_gen = input_generator(BATCH_SIZE)
37   validation_gen = input_generator(BATCH_SIZE)
38   model.fit_generator(input_gen,
39                  epochs = 10, workers=1,
40                  steps_per_epoch = 1000,
41                  validation_data = validation_gen,
42                  validation_steps = 10)
43
44 model = build_model()
45 model.summary()
46 train(model)
47
48 # ...main block...
```

In this program, the training data consists of pairs of input-output, where the input is one-hot encoded and the output is exactly the output of the function implemented by the hardcoded counterpart. The function encode_values (lines #5–15) does just that. The input_generator (lines #17–30) generates one-hot encoded inputs and vectors of real numbers as outputs (see lines #28 and #29).

Let's look at the train function (lines #32–42). In the previous chapter, training was done using the categorial cross-entropy loss function, because the output was one-hot representations. But in here, the output is a vector of real numbers. We know there is supposed to be only one 1, but that is not something we can express. We are dealing with learning an actual function from categorical one-hot inputs to a vector of real numbers. In machine learning terminology, the problem at hand is *regression* as opposed to *classification*. Regression is the problem of predicting *a continuous value* based on values seen before; classification is the problem of predicting *a category* based on values seen before. In our case, given that the output is the set of real numbers we want to predict, we use the simple loss function *mean square error* (MSE) (line #33). We also use more training data than in the previous chapter (200,000 as opposed to 4,000), because the function here is more complex, and takes a little longer to learn.

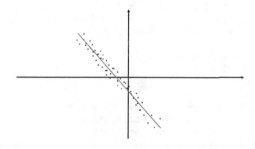

Linear regression is the statistical basis upon which machine learning works, including classification. Invented in the beginning of the 19th century, linear regression is a family of techniques for finding the straight line that best fits a set of data points (see figure above). Essentially, supervised machine learning is about learning a function given a training set of examples (data points). Of course, there is a lot more to supervised machine learning than linear regression. Specifically, the derived function should be *generalizable*, in the sense that it should work well for data points that are not part of the training set. But linear regression is at the core of learning functions from data.

This learning program, however, is a bit disappointing. We are making the network learn a very narrow function that transforms characters into a strange vector representation, and then we have to decode that representation using our decode_values function. And while this seemed natural in the hardcoded version of the neural network, having to decode this strange vector by hand in the learning version seems much ado about nothing. Can we make the network also learn how to decode the vector into the nice one-hot representation of characters?

Well, yes, we can. In fact, we could have also made the hardcoded network do that, too. One approach would be thresholding by zooming in on the 1. But thresholding does not work well with backpropagation. Another approach is to add another dense layer that learns how to translate the strange vector to one-hot encoded representations, like so:

```python
def build_model():
    model = Sequential()
    model.add(Dense(1, input_shape=(INPUT_VOCAB_SIZE,)))
    model.add(Dense(INPUT_VOCAB_SIZE))
    model.add(Dense(INPUT_VOCAB_SIZE, activation='softmax'))
    return model

def train(model):
    model.compile(loss='categorical_crossentropy',
                  optimizer='adam',
                  metrics=['accuracy'])
    input_gen = input_generator(BATCH_SIZE)
    validation_gen = input_generator(BATCH_SIZE)
    model.fit_generator(input_gen,
                epochs = 10, workers=1,
                steps_per_epoch = 1000,
                validation_data = validation_gen,
                validation_steps = 10)

def input_generator(nsamples):
    def generate_line():
        # ...same...
    while True:
        input_data, expected = generate_line()
        data_in = encode_one_hot(input_data)
        data_out = encode_one_hot(expected)
        yield data_in, data_out
```

Noteworthy here is the extra layer in line #5. This is the layer that takes the vector of real numbers and produces one-hot representations – or so we hope. Because of this layer, we can now train the network again on one-hot inputs to one-hot outputs, so input_generator is yielding one-hot outputs again (line #26). Finally, since we are back to using categories, training uses the categorical cross-entropy loss function again.

In summary: the single hidden neuron preserves the information but destroys the feature independence of the input, by transforming it into one real-valued number; the decoding layer recovers the main coding property of the output – the fact that only one of the neurons carries all of, and only, the input signal; finally, the last layer is able to capture that main property and transform it into one-hot representation. The data is recovered in categorical format!

The drawback of this second model is that it adds 10,000 weights – a heavy hand, considering that the original network is just 200.

37.4   HISTORICAL NOTES

The term *deep learning* is associated to neural networks that have more than one hidden layer, such as the final one in this chapter. In contrast, both the initial bow tie network in this chapter and the neural networks in the two previous chapters are called *shallow* networks. While the field of neural networks aimed, from the beginning, to tackle all sorts of neural networks, the learning techniques originally proposed did not work for multi-layer networks. Without multi-layer networks, the applicability of neural networks was quite limited. This was a stumbling block for the field for several decades.

In 1986, Rumelhart, Hinton and Williams (RHW) published a short, but highly influential paper in *Nature* explaining how to backpropagate errors in multi-layer neural networks. In that paper, they showed a few applications of their technique to solving complex classification problems. That paper marked the beginning of modern deep learning, i.e. learning in multi-layer networks.

The RHW paper built on techniques that had been around for a few years. Backpropagation itself was invented by multiple people in the 1960s, and had at least one known implementation in 1970 by Seppo Linnainmaa. In his PhD thesis in 1974, Paul Werbos showed how backpropagation could be used in multi-layer neural networks. But due to the "AI Winter" of the 1970s – a period of deep skepticism about AI – these works only came to light a decade later.

37.5   FURTHER READING

Linnainmaa, S. (1970). The Representation of the Cumulative Rounding Error of an Algorithm as a Taylor Expansion of the Local Rounding Errors. Master's thesis, Univ. Helsinki.
Synopsis: The first known implementation of backpropagation.

Rumelhart, D.E., Hinton, G.E. and Williams, R.J. (1986). Learning representations by back-propagating errors. *Nature*, 323(9):533–536.
Synopsis: Highly influential paper that popularized the concept of backpropagation in multi-layer neural networks. The idea of backpropagation had been around since the 1960s, and was independently discovered by several people, in several fields, some of which were unrelated to neural networks.

P. Werbos (1974). Beyond Regression: New Tools for Prediction and Analysis in the Behavioral Sciences. PhD thesis, Harvard University, Cambridge, MA.
Synopsis: One of the first applications of backpropagation to multi-layer neural networks.

37.6   GLOSSARY

Backpropagation: Analytical solution to calculating the derivative of the error as a function of the weights in neural networks.

Deep learning: Supervised learning using neural networks with more than one hidden layer.

Gradient descent: Family of optimization algorithms for minimizing errors. Gradient descent does not have analytical solutions, so these algorithms are iterative.

Hidden layer: A layer that is neither the input nor the output of a neural network.

37.7   EXERCISES

37.1 *Manual.* Implement the categorical version of the problem (the last code snippet) by hardcoding the weights of the third (and last) layer.

37.2 *Another encoding.* Implement the example program, using a bow tie network, using ACII encoding instead of one-hot encoding.

37.3 *Another function.* Implement an NN that learns to transform characters into their LEET counterparts. Use any encoding you want and any LEET code you want.

## CHAPTER 38

Neuro-Monolithic

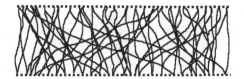

38.1 CONSTRAINTS

▷ Many conceptually different functions are implemented in one single dense layer.

▷ Certain outputs that aren't logically related to certain inputs are made artificially related.

38.2   A PROGRAM IN THIS STYLE

```
1  from keras.models import Sequential
2  from keras.layers import Dense
3  import numpy as np
4  import sys, os, string
5
6  characters = string.printable
7  char_indices = dict((c, i) for i, c in enumerate(characters))
8  indices_char = dict((i, c) for i, c in enumerate(characters))
9
10 INPUT_VOCAB_SIZE = len(characters)
11 LINE_SIZE = 80
12
13 def encode_one_hot(line):
14     x = np.zeros((1, LINE_SIZE, INPUT_VOCAB_SIZE))
15     sp_idx = char_indices[' ']
16     for i, c in enumerate(line):
17         index = char_indices[c] if c in characters else sp_idx
18         x[0][i][index] = 1
19     # Pad with spaces
20     for i in range(len(line), LINE_SIZE):
21         x[0][i][sp_idx] = 1
22     return x.reshape([1, LINE_SIZE*INPUT_VOCAB_SIZE])
23
24 def decode_one_hot(y):
25     s = []
26     x = y.reshape([1, LINE_SIZE, INPUT_VOCAB_SIZE])
27     for onehot in x[0]:
28         one_index = np.argmax(onehot)
29         s.append(indices_char[one_index])
30     return ''.join(s)
31
32 def normalization_layer_set_weights(n_layer):
33     wb = []
34     w = np.zeros((LINE_SIZE*INPUT_VOCAB_SIZE, LINE_SIZE*
           INPUT_VOCAB_SIZE))
35     b = np.zeros((LINE_SIZE*INPUT_VOCAB_SIZE))
36     for r in range(0, LINE_SIZE*INPUT_VOCAB_SIZE, INPUT_VOCAB_SIZE
           ):
37         # Let lower case letters go through
38         for c in string.ascii_lowercase:
39             i = char_indices[c]
40             w[r+i, r+i] = 1
41         # Map capitals to lower case
42         for c in string.ascii_uppercase:
43             i = char_indices[c]
44             il = char_indices[c.lower()]
45             w[r+i, r+il] = 1
46         # Map all non-letters to space
47         sp_idx = char_indices[' ']
48         for c in [c for c in list(string.printable) if c not in
               list(string.ascii_letters)]:
49             i = char_indices[c]
50             w[r+i, r+sp_idx] = 1
51         # Map single letters to space
```

```
52          previous_c = r-INPUT_VOCAB_SIZE
53          next_c = r+INPUT_VOCAB_SIZE
54          for c in [c for c in list(string.printable) if c not in
               list(string.ascii_letters)]:
55              i = char_indices[c]
56              if r > 0 and r < (LINE_SIZE-1)*INPUT_VOCAB_SIZE:
57                  w[previous_c+i, r+sp_idx] = 0.75
58                  w[next_c+i, r+sp_idx] = 0.75
59              if r == 0:
60                  w[next_c+i, r+sp_idx] = 1.5
61              if r == (LINE_SIZE-1)*INPUT_VOCAB_SIZE:
62                  w[previous_c+i, r+sp_idx] = 1.5
63
64      wb.append(w)
65      wb.append(b)
66      n_layer.set_weights(wb)
67      return n_layer
68
69 def build_model():
70      # Normalize characters using a dense layer
71      model = Sequential()
72      model.add(Dense(LINE_SIZE*INPUT_VOCAB_SIZE,
73                  input_shape=(LINE_SIZE*INPUT_VOCAB_SIZE,),
74                  activation='sigmoid'))
75      return model
76
77 model = build_model()
78 model.summary()
79 normalization_layer_set_weights(model.layers[0])
80
81 with open(sys.argv[1]) as f:
82      for line in f:
83          if line.isspace(): continue
84          batch = encode_one_hot(line)
85          preds = model.predict(batch)
86          normal = decode_one_hot(preds)
87          print(normal)
```

38.3 COMMENTARY

THE previous two chapters focused on the first part of the term frequency problem: normalization of the characters. Let us now turn our attention to the second part of the problem: eliminating single-letter words. When a single letter is between two spaces, it should be replaced by a white space, as a form of elimination. In order to do that, we cannot look at characters in isolation; we need to look, at least, at the characters before and after any given character.

Capturing dependencies between a *sequence* of inputs in neural networks is, perhaps, one of the most fascinating aspects of NNs, because it leads to reevaluating our understanding of both *space* (i.e. storage, memory) and *time* in computing. The simple *feed forward* neural networks we have seen so far are stateless machines – they do not have the capability of storing information about previous inputs. In the next few chapters, we will see how memory of past inputs can emerge in NNs. But to start with, in here we adhere to the strict constraint of a feed forward, stateless dense layer.

A first way of approaching the problem is to directly trade time with space, in a one-to-one manner. That is, rather than processing one character at a time, let's process an entire line at a time. That way, we can "program" not only the character transformation functions we had before, but also the dependencies among the characters at different positions of the line. We will have a much larger input and output, and a quadratic increase on the number of connections.

The example program does exactly that. There are many similarities between this program and the one in Chapter 35. Here, too, the network is "programmed" not by learning but by manually setting the weights, just so it is clear what the logic of character dependencies can look like in the connectionist style. Both programs have the same functions, and very similar implementations of them. Let's focus on the differences.

One of the main differences is that we now need to establish the maximum size of a line, because the dense layer processes the entire line – all input to neural networks requires fixed-size tensors. The size of the line is defined in line #11 LINE_SIZE = 80. If the given line is less than the maximum size, the one-hot encoding function pads it with space characters (lines #20-21).

Another size-related detail is an extra first dimension of the input – see, for example, lines #14, #18, #21, and #22. In the previous two chapters, this didn't need explanation, because it corresponded naturally to the number of characters of each line; but here it is not so obvious. Keras requires all input to have at least 2 dimensions: units of the input data are given to the network in fixed-size collections called *batches*. As such, the first dimension of the input is always the batch. When there is always only one piece of data, such as the case of the predictions here (line by line), then the batch dimension is 1.

The dense layer of the model (lines #72–74) now maps input of size LINE_SIZE * INPUT_VOCAB_SIZE = 8,000 to output of the same size. That means that the number of connections is a whopping 64 million! (Compare to 10,000 of Chapter 35.) Dense layers quickly become unscalable for large input sizes. As in Chapter 35, the vast majority of these connections have zero weight; only a couple of hundred are not zero.

Let's look at our network "program" given in lines #32–67. The processing of lowercase letters (lines #38–40), uppercase letters (lines #42–45), and non-letter characters (lines #47–50) is exactly the same as before. Our "program" now includes a few more connections to deal with inter-character dependencies in lines #52–62. The idea is the following: for every character of the string, we look at the characters before and after; for any of those characters that are not a letter, we add a non-zero weight from their 1-position to the space index of the output of the current character. The exact value of those weights is not important, but it needs to obey two rules: one weight should be less than 1; and the addition of two weights should be greater than 1. That way, when the current character is surrounded by non-letters, it receives two weights from the neighboring inputs, as if votes, to map it to the space character. If it only has one non-letter as a neighbor, then it only receives one weight for the space character, which is not enough for the output to be interpreted as the space.

Note that, just like in the bow tie example of the previous chapter, this results in character outputs that are no longer one-hot encoded, but that can have more than one non-zero value. This is not a problem for decoding the vectors back into characters: our interpretation of which position the 1 is (function decode_one_hot, lines #24–30) finds the position of the maximum value, not the position of the exact value 1 (line #28).

Can this network be automatically programmed by learning on training data? The answer is yes, it can. The following snippet transforms the example program into its learning counterpart:

```
 1 BATCH_SIZE = 200
 2 STEPS_PER_EPOCH = 5000
 3 EPOCHS = 4
 4
 5 # ...Encoding functions...
 6
 7 def input_generator(nsamples):
 8     def generate_line():
 9         inline = []; outline = []
10         for _ in range(LINE_SIZE):
11             c = random.choice(characters)
12             expected = c.lower() if c in string.ascii_letters else
                         ' '
13             inline.append(c); outline.append(expected)
14         for i in range(LINE_SIZE):
15             if outline[i] == ' ': continue
16             if i > 0 and i < LINE_SIZE - 1:
17                 outline[i] = ' ' if outline[i-1] == ' ' and
                             outline[i+1] == ' ' else outline[i]
18             if (i == 0 and outline[i+1] == ' ') or (i == LINE_SIZE
                             -1 and outline[i-1] == ' '):
19                 outline[i] = ' '
```

```
20          return ''.join(inline), ''.join(outline)
21
22      while True:
23          data_in = np.zeros((nsamples, LINE_SIZE*INPUT_VOCAB_SIZE))
24          data_out = np.zeros((nsamples, LINE_SIZE*INPUT_VOCAB_SIZE)
                )
25          for i in range(nsamples):
26              input_data, expected = generate_line()
27              data_in[i] = encode_one_hot(input_data)[0]
28              data_out[i] = encode_one_hot(expected)[0]
29          yield data_in, data_out
30
31  def train(model):
32      model.compile(loss='binary_crossentropy',
33                    optimizer='adam',
34                    metrics=['accuracy'])
35      input_gen = input_generator(BATCH_SIZE)
36      validation_gen = input_generator(BATCH_SIZE)
37      model.fit_generator(input_gen,
38                  epochs = EPOCHS, workers=1,
39                  steps_per_epoch = STEPS_PER_EPOCH,
40                  validation_data = validation_gen,
41                  validation_steps = 10)
42
43  model = build_deep_model()
44  model.summary()
45  train(model)
46
47  # Main block below
```

Note that now the loss function (line #32) is binary, not categorical, cross-entropy. In Chapter 36 we used categorical cross-entropy. The difference is the following. In Chapter 36 the output was a one-hot encoded representation of a character, and therefore, we wanted the NN to learn to distinguish the exact "category" (i.e. the exact one-encoding neuron) from all the others. Hence categorical cross-entropy, which takes into account a set of output neurons. But here, the output is a set of 8,000 neurons, 80 of which are 1, the others 0. The complete set of outputs is not one-hot representation anymore. As such, we want the NN to learn to distinguish 0s from 1s, independently, for all output neurons. Hence, binary cross-entropy.[1]

However, training this massive 64M-weight network is not easy. As a rule of thumb, the more trainable weights an NN has, the more training data it requires to learn. The heuristics for this rule of thumb vary, because they depend on the problem, but roughly, we need at least an amount of data close to the same order of magnitude of the amount of weights. For the previous character-to-character NN, we had 10,000 weights and 4,000 data samples in the training set. With 64M weights, we need around 1M samples. The constants in the snippet above (lines #1–3) establish that number (remember,

---

[1]In machine learning terminology, we are dealing with a *multi-label classification* problem.

size of training set = steps per epoch × batch size). With 1M samples and 64M-weights, it takes a long time to train this network – mileage may vary, depending on whether GPUs are being used or not. The reader is encouraged to experiment with the training parameters, especially the size of the training set (as determined by STEPS_PER_EPOCH).

Our massive network in this program is an example of monolithic thinking in the connectionist world. We just want to solve the problem without thinking much about it, and by overloading all the logic in one single layer. This way of thinking has some similarities with the style of Chapter 4:

> From a design perspective, the main concern is to obtain the desired output without having to think much about subdividing the problem or how to take advantage of code that already exists. Given that the entire problem is one single conceptual unit, the programming task consists of defining the data and control flow that rule this unit.

In the connectionist model, the "control flow" is embodied in the connections between neurons, and the values of the weights. In this style, we load all of the logic in one single massive dense layer, and hope for the best.

Chapter 4 also introduced the concept of *cyclomatic complexity* as a proxy metric for program understandability. Neural networks have an equivalent metric: the number of *trainable parameters*, i.e. the number of connections that can be programmed either manually or with a learning algorithm. The higher the number of trainable parameters, the harder it will be to program/train them. Clearly, 64 million trainable weights is a huge number, and it is totally unjustified for this simple problem. Thinking a bit harder about the nature of the problem, and how to represent temporal data, will lead to much smaller and modular solutions, which will be much easier to train and understand.

38.4   GLOSSARY

Feed Forward: A neural network without cycles.

Trainable parameters: The weights and biases of a neural network that are updated during backpropagation.

38.5   EXERCISES

38.1   *Prove it.* There are many other solutions to the problem at hand, with different weights. Can you find a solution using neuro monolithic style that preserves one-hot encoding of the output? If so, show it. If not, prove that no solution exists that ensures one-hot encodings on the output.

38.2   *More eliminations.* Change the example program so that it also transforms the first and last characters of a line into some other character (e.g. the space) if they are the same.

## CHAPTER 39

Sliding Window

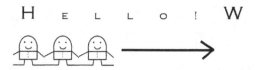

39.1 CONSTRAINTS

▷ The input is a sequence of items, and the output depends on certain patterns in that sequence.

▷ The input is reshaped as a sequence of concatenations of N items in the original sequence, where N is large enough to be able to capture the pattern.

▷ The concatenations are created by sliding through the input sequence, with a step of S, depending on the problem at hand.

39.2   A PROGRAM IN THIS STYLE

```
1  from keras.models import Sequential
2  from keras.layers import Dense
3  import numpy as np
4  import sys, os, string
5
6  characters = string.printable
7  char_indices = dict((c, i) for i, c in enumerate(characters))
8  indices_char = dict((i, c) for i, c in enumerate(characters))
9
10 INPUT_VOCAB_SIZE = len(characters)
11 WINDOW_SIZE = 3
12
13 def encode_one_hot(line):
14     line = " " + line + " "
15     x = np.zeros((len(line), INPUT_VOCAB_SIZE))
16     for i, c in enumerate(line):
17         index = char_indices[c] if c in characters else
             char_indices[' ']
18         x[i][index] = 1
19     return x
20
21 def decode_one_hot(x):
22     s = []
23     for onehot in x:
24         one_index = np.argmax(onehot)
25         s.append(indices_char[one_index])
26     return ''.join(s)
27
28 def prepare_for_window(x):
29     # All slices of size WINDOW_SIZE, sliding through x
30     ind = [np.array(np.arange(i, i+WINDOW_SIZE)) for i in range(x.
         shape[0] - WINDOW_SIZE + 1)]
31     ind = np.array(ind, dtype=np.int32)
32     x_window = x[ind]
33     # Reshape it back to a 2-d tensor
34     return x_window.reshape(x_window.shape[0], x_window.shape[1]*
         x_window.shape[2])
35
36 def normalization_layer_set_weights(n_layer):
37     wb = []
38     w = np.zeros((WINDOW_SIZE*INPUT_VOCAB_SIZE, INPUT_VOCAB_SIZE))
39     b = np.zeros((INPUT_VOCAB_SIZE))
40     # Let lower case letters go through
41     for c in string.ascii_lowercase:
42         i = char_indices[c]
43         w[INPUT_VOCAB_SIZE+i, i] = 1
44     # Map capitals to lower case
45     for c in string.ascii_uppercase:
46         i = char_indices[c]
47         il = char_indices[c.lower()]
48         w[INPUT_VOCAB_SIZE+i, il] = 1
49     # Map all non-letters to space
50     sp_idx = char_indices[' ']
```

```
51    non_letters = [c for c in list(characters) if c not in list(
          string.ascii_letters)]
52    for c in non_letters:
53        i = char_indices[c]
54        w[INPUT_VOCAB_SIZE+i, sp_idx] = 1
55    # Map single letters to space
56    for c in non_letters:
57        i = char_indices[c]
58        w[i, sp_idx] = 0.75
59        w[INPUT_VOCAB_SIZE*2+i, sp_idx] = 0.75
60
61    wb.append(w)
62    wb.append(b)
63    n_layer.set_weights(wb)
64    return n_layer
65
66 def build_model():
67    # Normalize characters using a dense layer
68    model = Sequential()
69    model.add(Dense(INPUT_VOCAB_SIZE,
70                    input_shape=(WINDOW_SIZE*INPUT_VOCAB_SIZE,),
71                    activation='softmax'))
72    return model
73
74 model = build_model()
75 model.summary()
76 normalization_layer_set_weights(model.layers[0])
77
78 with open(sys.argv[1]) as f:
79    for line in f:
80        if line.isspace(): continue
81        batch = prepare_for_window(encode_one_hot(line))
82        preds = model.predict(batch)
83        normal = decode_one_hot(preds)
84        print(normal)
```

39.3 COMMENTARY

THE PROBLEM of eliminating single-letter words from the input stream does not require knowledge of the entire line; it can be solved by looking at just three consecutive characters at a time. We can, therefore, design a network that slides through the input line, 3 characters at a time, and outputs the correct character. The example program does that. Let's analyze it.

As before, characters are encoded using one-hot representation. Since there are 100 printable characters, we are dealing with tensors of size 100 for each character. With three characters as input, we have a network that receives 300 inputs (line #70) and produces one single character of size 100 (line #69). This network is much smaller than the previous one: only 30,000 neural connections: 300 inputs to 100 outputs. Before analyzing the input, let's focus on the logic of the connections.

Just like the previous chapter, the example program here manually "programs" the network by setting the weights. That function is in lines #36–64. The logic is the same as the previous chapter. The only difference is that the weight matrix is only $300 \times 100 = 30,000$ ($WINDOW\_SIZE * INPUT\_VOCAB\_SIZE \times INPUT\_VOCAB\_SIZE$).

Perhaps the most important part of thinking about the problem in this sliding-window style is the setup of the input. In fact, setting up the input for NNs – both the encoding and the shape – is an important part of solving the problem! More on this later. Let's look at what the example program does. The function for encoding characters with one-hot representation (lines #13–19) is identical to the previous chapters, with one small difference: we are adding an extra space to the beginning and to the end of every line (line #14). This is the same trick done in Chapter 3 (Array Programming Style), as it simplifies the logic for dealing with the first and last characters of the given line.

Having added those extra spaces at the edges, the question now is how to prepare the input for the network. Suppose, for example, that the line is "I am a dog!" For clarity, let's rewrite this line using a visible character as replacement for the white space, and including the extra white space on the edges: ⌣I⌣am⌣a⌣dog!⌣ A possible, but wrong, input would be: [⌣I⌣], [am⌣], [a⌣d], [og!] [⌣⌣⌣]. This would be wrong for many reasons: first, we would only be producing one output character for every three input characters; then, the network would miss the single-letter 'a' word. Instead, we need a sliding window that produces the following sequence: [⌣I⌣], [I⌣a], [⌣am], [am⌣], [m⌣a], [⌣a⌣], [a⌣d], [⌣do], [dog], [og!], [g!⌣]. By sliding through every character of the line, one by one, the network produces the same number of characters on the output, while making its decision about the middle character of each triplet, based not only on the character itself, but also on the neighboring two characters.

To this end, the example program includes a new function in lines #28–34, `prepare_for_window`, which reshapes the sequence of one-hot-encoded characters x into the proper sequence of triplets. It does so by using array programming style operations. In line #30, we generate all start:end indices for all the triplets of the input, and create an array with all of them (line #31). That array is then used to slice the input array, all at once (line #32). The slicing operation produces a 3D tensor with the shape (len(line), WINDOW_SIZE, INPUT_VOCAB_SIZE); before we feed it to the network, we need to reshape it for 2 dimensions only: one sequence of triplets. That is done in line #34.

39.4   GLOSSARY

Reshape: To rearrange the data of a multidimensional tensor so that it fits different dimensions. For example, a 1-dimensional array of size 100 to a 2-dimensional array of size 10×10.

39.5   EXERCISES

39.1   *Train.* Train the network of the example program instead of hardcoding the weights. Note that you need to pay attention to the encoding of the output.

39.2   *More eliminations.* Change the example program so that it also eliminates 2-letter words.

## CHAPTER 40

Recurrent

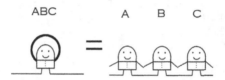

40.1 CONSTRAINTS

▷ The input is a sequence of items, and the output depends on certain patterns in that sequence.

▷ The length of the output sequence is exactly the same as the length of the input sequence.

▷ The input is reshaped as a sequence of frames of size N, each one capturing a portion of the input sequence, large enough to be able to capture the pattern.

▷ The frames are created by sliding through the input sequence, with a step of 1.

▷ The neural function is defined as a single unit that is instantiated N times and applied to all items in a frame, at the same time. Those instances are connected in a chain where the output of one feeds as input into the next. Each unit has, therefore, two sets of weights: one for the items from the input sequence, the other for the output of the previous unit in the chain.

40.2   A PROGRAM IN THIS STYLE

```
1  from keras.models import Sequential
2  from keras.layers import Dense, SimpleRNN
3  import numpy as np
4  import sys, os, string, random
5
6  characters = string.printable
7  char_indices = dict((c, i) for i, c in enumerate(characters))
8  indices_char = dict((i, c) for i, c in enumerate(characters))
9
10 INPUT_VOCAB_SIZE = len(characters)
11 BATCH_SIZE = 200
12 HIDDEN_SIZE = 100
13 TIME_STEPS = 3
14
15 def encode_one_hot(line):
16     x = np.zeros((len(line), INPUT_VOCAB_SIZE))
17     for i, c in enumerate(line):
18         index = char_indices[c] if c in characters else
                  char_indices[' ']
19         x[i][index] = 1
20     return x
21
22 def decode_one_hot(x):
23     s = []
24     for onehot in x:
25         one_index = np.argmax(onehot)
26         s.append(indices_char[one_index])
27     return ''.join(s)
28
29 def prepare_for_rnn(x):
30     # All slices of size TIME_STEPS, sliding through x
31     ind = [np.array(np.arange(i, i+TIME_STEPS)) for i in range(x.
              shape[0] - TIME_STEPS + 1)]
32     ind = np.array(ind, dtype=np.int32)
33     x_rnn = x[ind]
34     return x_rnn
35
36 def input_generator(nsamples):
37     def generate_line():
38         inline = [' ']; outline = []
39         for _ in range(nsamples):
40             c = random.choice(characters)
41             expected = c.lower() if c in string.ascii_letters else
                       ' '
42             inline.append(c); outline.append(expected)
43         inline.append(' ');
44         for i in range(nsamples):
45             if outline[i] == ' ': continue
46             if i > 0 and i < nsamples-1:
47                 if outline[i-1] == ' ' and outline[i+1] == ' ':
48                     outline[i] = ' '
49             if (i == 0 and outline[1] == ' ') or (i == nsamples-1
                   and outline[nsamples-2] == ' '):
50                 outline[i] = ' '
```

```
51            return ''.join(inline), ''.join(outline)
52
53     while True:
54         input_data, expected = generate_line()
55         data_in = encode_one_hot(input_data)
56         data_out = encode_one_hot(expected)
57         yield prepare_for_rnn(data_in), data_out
58
59 def train(model):
60     model.compile(loss='categorical_crossentropy',
61                   optimizer='adam',
62                   metrics=['accuracy'])
63     input_gen = input_generator(BATCH_SIZE)
64     validation_gen = input_generator(BATCH_SIZE)
65     model.fit_generator(input_gen,
66                 epochs = 50, workers=1,
67                 steps_per_epoch = 50,
68                 validation_data = validation_gen,
69                 validation_steps = 10)
70
71 def build_model():
72     model = Sequential()
73     model.add(SimpleRNN(HIDDEN_SIZE, input_shape=(None,
               INPUT_VOCAB_SIZE)))
74     model.add(Dense(INPUT_VOCAB_SIZE, activation='softmax'))
75     return model
76
77 model = build_model()
78 model.summary()
79 train(model)
80
81 input("Network has been trained. Press <Enter> to run program.")
82 with open(sys.argv[1]) as f:
83     for line in f:
84         if line.isspace(): continue
85         batch = prepare_for_rnn(encode_one_hot(line))
86         preds = model.predict(batch)
87         normal = decode_one_hot(preds)
88         print(normal)
```

40.3 COMMENTARY

THE SLIDING WINDOW style of the previous chapter is a special-purpose, highly optimized solution to the problem at hand: the elimination of single-letter words. In order to capture dependencies on sequences of inputs, in general, a special kind of neural network is typically used: a *recurrent* neural network (RNN). Conceptually, a recurrent layer is illustrated by the figure below:

input

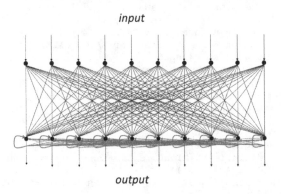

output

The arrows at the bottom mean that the output of this layer depends not just on its input but also on the value of the output at an earlier time. Mapping this concept to our problem, it means that the exact output character depends on what characters were seen before. These kinds of circular dependencies are the bread and butter of traditional programming, and we have developed good conceptual tools for dealing with them – namely, iteration and recursion. Neural networks, however, challenge what we know about this, because there is neither control flow nor recursion. Those concepts need to be invented here, from scratch, and they take very different forms from the ones we are used to in traditional programming.

Before diving into the example program, let's bring a concept from traditional programming that helps us understand recurrent networks: *loop unrolling*. Loop unrolling is a technique that tries to optimize program execution by transforming loops into sequences of instruction blocks corresponding to the body of the loop. The idea is to eliminate the instructions that control the loop itself, such as the conditional and the increment. These optimizations aren't always possible, but, in certain circumstances, they can be – for example, when we know the exact number of iterations, we can simply copy-paste the loop block as many times, and make the necessary adjustments on the values of variables in each block.

A similar technique is used in RNNs: because there is no control flow, we cannot express the concept of a loop in the first place; loop unrolling is not just an optimization here, it's a necessity. In NNs, we can express the concept of a function that is applied N times, repeatedly, by adding N layers to the model, each one doing exactly the same thing. Like so:

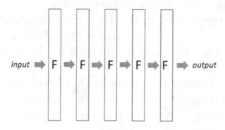

Although the example programs, so far, haven't used it, there are ways of expressing *weight sharing* between different layers of the network. If we hard-code it, sharing weights is trivial – we simply apply the same weights/bias matrices to several layers. In learning, weight sharing serves as a constraint during training: the different layers that share weights are guaranteed to have the same weights during backpropagation. Layers that share weights implement the exact same function.

But that is still not sufficient for expressing dependencies between different items of the input sequence. Yes, we need repetitions, and those can be expressed via weight sharing; but we also need access to information from within a *window* of the input sequence. That is done with a touch of ingenuity on how to think about discrete *time* – the sequence of inputs, to be precise – in NNs. The core of the idea is illustrated in the following figure:

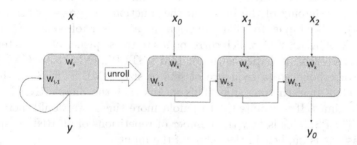

The figure is deceivingly simple, because it looks like loop unrolling in traditional programming, so it's easy to miss important details. Let's analyze it carefully. The first thing to notice is the repetition of the same unit, in this case, 3 times; that is weight sharing. The next thing to notice is that each unit has two sets of learnable parameters, not just one. One of the sets is the regular input-to-output weight/bias parameters that we have already seen.

In the figure, it is denoted as W_x. The second set of parameters (W_{t-1}) is the previous-output-to-output weight/bias parameters, i.e. the arrows at the bottom of the first figure in this chapter. Finally, there is one more important detail related to the input itself: a window of input items is spread along as input to the N repetitions; in this case that window is of size 3. *Only at the end of every 3 inputs do we get one output.* That is, $y_0 = f(x_0, x_1, x_2)$, $y_1 = f(x_1, x_2, x_3)$, etc. The output sequence is *shifted* with respect to the input sequence.

In summary, RNNs implement loops by (a) instantiating the same function N times; (b) connecting the instances of that function in a chain via their outputs; and (c) taking N items of the input sequence at once, and giving each item to each instance simultaneously.

One last thing remains unclear: how do we specify N, the number of repetitions? This is, perhaps, one of the most obscure elements of RNNs in Keras. Essentially, that important parameter is implicit in the *shape* of the input! – we will see how in the example program.

Having described the basic concepts in RNNs, let's finally analyze the example program, starting with the model definition in lines #71–75. This model consists of a SimpleRNN layer that takes input in some shape where the last dimension is INPUT_VOCAB_SIZE (100), and produces output of the same size, 100. The output of that RNN is then fed to a Dense layer with a softmax activation. We are applying here the knowledge covered in Chapter 37, namely the idea of applying one softmax-activated dense layer after a regression layer, to recover the categories in one-hot representation. The program is presented in its learning version, so the training part is there (lines #59–69). The training configuration is the one expected for categorical learning, so there is nothing new in this function.

The second noteworthy, and hard to understand, element of this program is the shaping of the input, in the function `prepare_for_rnn` (lines #29–34). The input to that function is a tensor of shape ((len(line), INPUT_VOCAB_SIZE), which corresponds to the sequence of characters. The output is a tensor of shape (len(line), TIME_STEPS, INPUT_VOCAB_SIZE), which corresponds to a sequence of sets of size TIME_STEPS, in this case 3 characters in each time step. This means that the SimpleRNN layer will unfold 3 times. If we wanted it to unfold more times, we would increase the TIME_STEPS. This is how the number of repetitions of an RNN is specified in Keras – it is implicit in the shape of the input.

The final noteworthy element is the generation of the sequence of inputs-outputs for training, in lines #36–57, in particular the inner function `generate_line`, which now does something a bit odd. Specifically, that function is not just generating inputs and corresponding outputs, like before, but is also performing a shift between the input sequence and the output sequence. This is done in line #38, with the insertion of a space character in the beginning of the input sequence, without the corresponding space character in beginning of the output sequence. The input sequence is shifted by one with respect to the output sequence. Why? Recall the figure depicting the loop unrolling in page 325. We want to center each character x_i in the middle position so that we can decide what the output y_{i-1} should be. Shifting by one accomplishes that.

40.4 HISTORICAL NOTES

One of the earliest works studying neural networks "with cycles" that were able to store state over time was published by W. Little in 1974. A few years later, those ideas were popularized by John Hopefield, in what is now known as Hopefield networks. The general form of Recurrent Neural Networks was used in Rumelhart, Hinton and Williams' *Nature* paper in 1986.

40.5 FURTHER READING

Little, W.A. (1974). The existence of persistent states in the brain. *Mathematical Biosciences* 19(1-2): February.

> *Synopsis*: One of the first papers exploring how state could emerge in [recurrent] neural networks. The ideas here are known as Hopefield networks, after John Hopefield, who popularized them a few years later.

40.6 GLOSSARY

Loop unrolling: An optimization technique in traditional programming whereby loops are eliminated by explicitly copy-pasting their body several times.

Weight sharing: Weight sharing ≈ applying the same function.

40.7 EXERCISES

40.1 *Under control.* Implement the example program by hardcoding the weights of the several layers.

40.2 *More eliminations.* Change the example program so that it also eliminates 2-letter words.

40.3 *Another pattern.* Implement an RNN that learns to transform the pattern cc (i.e. two repeated characters) into 'xx'.

40.4 *Telephone numbers.* Implement an RNN that learns to anonymize telephone numbers. Assume that telephone numbers are given as any sequence of 11 digits (e.g. 9495551123), or a sequence of 11 digits with dashes (e.g. 949-5551123, 949-555-1123). When detected, all the digits in the pattern should be replaced by x (e.g. xxxxxxxxxx or xxx-xxxxxxx or xxx-xxx-xxxx, respectively).

40.5 *Stop words.* Implement an RNN that learns to replace all occurrences of stop words with spaces. Use the stop words given by the file stop_words.

Index

Printed in the United States
by Baker & Taylor Publisher Services